Springer Complexity

Springer Complexity is an interdisciplinary program publishing the best research and academic-level teaching on both fundamental and applied aspects of complex systems – cutting across all traditional disciplines of the natural and life sciences, engineering, economics, medicine, neuroscience, social and computer science.

Complex Systems are systems that comprise many interacting parts with the ability to generate a new quality of macroscopic collective behavior the manifestations of which are the spontaneous formation of distinctive temporal, spatial or functional structures. Models of such systems can be successfully mapped onto quite diverse "real-life" situations like the climate, the coherent emission of light from lasers, chemical reaction-diffusion systems, biological cellular networks, the dynamics of stock markets and of the internet, earthquake statistics and prediction, freeway traffic, the human brain, or the formation of opinions in social systems, to name just some of the popular applications.

Although their scope and methodologies overlap somewhat, one can distinguish the following main concepts and tools: self-organization, nonlinear dynamics, synergetics, turbulence, dynamical systems, catastrophes, instabilities, stochastic processes, chaos, graphs and networks, cellular automata, adaptive systems, genetic algorithms and computational intelligence.

The two major book publication platforms of the Springer Complexity program are the monograph series "Understanding Complex Systems" focusing on the various applications of complexity, and the "Springer Series in Synergetics", which is devoted to the quantitative theoretical and methodological foundations. In addition to the books in these two core series, the program also incorporates individual titles ranging from textbooks to major reference works.

Editorial and Programme Advisory Board

Péter Érdi
Center for Complex Systems Studies, Kalamazoo College, USA and Hungarian Academy of Sciences, Budapest, Hungary

Karl Friston
Institute of Cognitive Neuroscience, University College London, London, UK

Hermann Haken
Center of Synergetics, University of Stuttgart, Stuttgart, Germany

Janusz Kacprzyk
System Research, Polish Academy of Sciences, Warsaw, Poland

Scott Kelso
Center for Complex Systems and Brain Sciences, Florida Atlantic University, Boca Raton, USA

Jürgen Kurths
Nonlinear Dynamics Group, University of Potsdam, Potsdam, Germany

Linda Reichl
Center for Complex Quantum Systems, University of Texas, Austin, USA

Peter Schuster
Theoretical Chemistry and Structural Biology, University of Vienna, Vienna, Austria

Frank Schweitzer
System Design, ETH Zurich, Zurich, Switzerland

Didier Sornette
Entrepreneurial Risk, ETH Zurich, Zurich, Switzerland

Understanding Complex Systems

Founding Editor: J.A. Scott Kelso

Future scientific and technological developments in many fields will necessarily depend upon coming to grips with complex systems. Such systems are complex in both their composition – typically many different kinds of components interacting simultaneously and nonlinearly with each other and their environments on multiple levels – and in the rich diversity of behavior of which they are capable.

The Springer Series in Understanding Complex Systems series (UCS) promotes new strategies and paradigms for understanding and realizing applications of complex systems research in a wide variety of fields and endeavors. UCS is explicitly transdisciplinary. It has three main goals: First, to elaborate the concepts, methods and tools of complex systems at all levels of description and in all scientific fields, especially newly emerging areas within the life, social, behavioral, economic, neuro- and cognitive sciences (and derivatives thereof); second, to encourage novel applications of these ideas in various fields of engineering and computation such as robotics, nano-technology and informatics; third, to provide a single forum within which commonalities and differences in the workings of complex systems may be discerned, hence leading to deeper insight and understanding.

UCS will publish monographs, lecture notes and selected edited contributions aimed at communicating new findings to a large multidisciplinary audience.

Pushkin Kachroo · Sadeq J. Al-nasur ·
Sabiha Amin Wadoo · Apoorva Shende

Pedestrian Dynamics

Feedback Control of Crowd Evacuation

With 66 Figures and 2 Tables

Dr. Pushkin Kachroo, P.E.
(Visiting) Associate Professor
Department of Electrical
and Computer Engineering
University of Nevada, Las Vegas
4505 S. Maryland Pkwy
Las Vegas, NV 89154-4026
pushkin@unlv.edu

Dr. Sabiha Amin Wadoo
Assistant Professor
Department of Electrical &
Computer Engineering
New York Institute of Technology
Northern Blvd, Old Westbury, NY 11568
swadoo@nyit.edu

Sadeq J. Al-nasur
Department of Electrical Engineering
Kuwait University
P. O. Box 5969
Safat, 13060 Kuwait
alnasur@eng.kuniv.edu.kw

Apoorva Shende
Bradley Dept. of Electrical &
Computer Engineering
Virginia Tech.
Blacksburg, VA 24060
U.S.A.
apoorva@vt.edu

ISBN: 978-3-540-75559-3   e-ISBN: 978-3-540-75561-6

Understanding Complex Systems ISSN: 1860-0832

Library of Congress Control Number: 2007938322

© 2008 Springer-Verlag Berlin Heidelberg

This work is subject to copyright. All rights are reserved, whether the whole or part of the material is concerned, specifically the rights of translation, reprinting, reuse of illustrations, recitation, broadcasting, reproduction on microfilm or in any other way, and storage in data banks. Duplication of this publication or parts thereof is permitted only under the provisions of the German Copyright Law of September 9, 1965, in its current version, and permission for use must always be obtained from Springer. Violations are liable to prosecution under the German Copyright Law.

The use of general descriptive names, registered names, trademarks, etc. in this publication does not imply, even in the absence of a specific statement, that such names are exempt from the relevant protective laws and regulations and therefore free for general use.

Cover design: Kirchner, Erich

Printed on acid-free paper

9 8 7 6 5 4 3 2 1

springer.com

To the Pursuit of Knowledge

Preface

The importance of controlling pedestrian flow especially during emergencies is being understood by researchers to be a very important research area. Currently, the use of static emergency routes is not efficient, since during emergencies, the preferred routes might be congested, or worse yet might not even exist. Hence, it is very important to use sensors to measure the current traffic and conditions on the routes and give real-time guidance to pedestrians using feedback control. This book is the first book that provides feedback control design for pedestrian movement control in one and two-dimensional problems using lumped and distributed parameter model settings. There is much more development that is needed in this important work, but the authors hope that this book provides inspiration for other researchers to continue work in this area.

Evacuation can be from a small area, single floor of a building, a entire building, a parking area, or from a much bigger region such as an entire city. The feedback control design for evacation of pedestrians in small areas falls under the framework presented in this book. Evacuation from bigger regions such as a city requires vehicular traffic control from highways, which can involve modeling of networks using digraphs. Network control for evacuation is not covered in this book.

This book is the outcome of research carried out for two years partially supported by National Science Foundation through grant no. CMS-0428196 with Dr. S. C. Liu as the Program Director. This support is gratefully acknowledged. Any opinion, findings, and conclusions or recommendations expressed in this study are those of the writer and do not necessarily reflect the views of the National Science Foundation. The research was conducted by the four authors of this book as well as Dr. M. P. Singh from Virginia Tech. Some earlier work also involved development of visualization software (not presented in this book) by Thomas A. Merrell.

Virginia Tech., (visiting) UNLV, *Pushkin Kachroo*
August 2007

Contents

1	**Introduction**	**1**
2	**Traffic Flow Theory for 1-D**	**5**
	2.1 Introduction	5
	2.2 Microscopic vs Macroscopic	6
	2.3 Car-Following Model	7
	2.4 Traffic Flow Theory	8
	2.4.1 Flow	8
	2.4.2 Conservation Law	9
	2.4.3 Velocity–Density Relationship(s)	11
	2.5 Traffic Flow Model 1-D	13
	2.5.1 LWR Model	14
	2.5.2 PW Model	15
	2.5.3 AR Model	17
	2.5.4 Zhang Model	19
	2.5.5 Models Summary	22
	2.6 Method of Characteristics	23
	2.6.1 LWR Model Classification	23
	2.6.2 Exact Solution	23
	2.6.3 Blowup of Smooth Solutions	25
	2.6.4 Weak Solution	27
3	**Crowd Models for 2-D**	**33**
	3.1 Introduction	33
	3.2 Traffic Flow Theory in 2-D	34
	3.3 One Equation Crowd Model	35
	3.4 First System Crowd Dynamic Model	36
	3.4.1 Model Description	36
	3.4.2 Conservation Form and Eigenvalues	38

- 3.5 Second Crowd Dynamic Model 40
  - 3.5.1 Model Description 40
  - 3.5.2 Conservation Form and Eigenvalues 41
- 3.6 Third Crowd Dynamic Model 44
  - 3.6.1 Model Description 44
  - 3.6.2 Derivation of a Macroscopic Model from a Microscopic Model in 2-D 45
  - 3.6.3 Conservation Form and Eigenvalues 46
- 3.7 Comparison Between the Models 50
- 3.8 Linearization 53
  - 3.8.1 One Equation Crowd Model 53
  - 3.8.2 First System Model 55
  - 3.8.3 Second System Model 56
  - 3.8.4 Third System Model 58

4 Numerical Methods   61
- 4.1 Introduction 61
- 4.2 Fundamentals of FVM 62
  - 4.2.1 Formulation of 2-D Numerical Schemes 63
- 4.3 Numerical Schemes 65
  - 4.3.1 Lax-Friedrichs Scheme 65
  - 4.3.2 FORCE Scheme 65
  - 4.3.3 Roe's Scheme 66
- 4.4 Simulation 67
  - 4.4.1 Initial and Boundary Conditions 67
  - 4.4.2 Simulation Results 68
- 4.5 Matlab Program Code 78
  - 4.5.1 One-equation Model 78
  - 4.5.2 First System Model 81
  - 4.5.3 Second System Model 87
  - 4.5.4 Third System Model 91

5 Feedback Linearization (1-D Patches)   95
- 5.1 Introduction 95
- 5.2 Theory 96
  - 5.2.1 Control Problem 96
  - 5.2.2 Characteristic Index 97
  - 5.2.3 State Feedback Control 97
  - 5.2.4 Closed-Loop Stability 98

	5.3	Application to the LWR (One patch) 98
	5.4	Application to the LWR ($n=5$ patches) 101
	5.5	Matlab Program Code 103
		5.5.1 One-patch Control 103
		5.5.2 Five-patch Control 104

6 Intelligent Evacuation Systems 107
- 6.1 Introduction 107
- 6.2 IPES Functions 109
- 6.3 IES Functions for Evacuation Scenarios 111
  - 6.3.1 Subway Station 111
  - 6.3.2 Airport 112
- 6.4 Four-Layer System Architecture 115
- 6.5 Four-Layer System: Scenarios 117
  - 6.5.1 Subway Station 117
  - 6.5.2 Airport 118
- 6.6 IT Issues and Requirements 118
- 6.7 Feedback Control and Dynamic Modeling 119

7 Discretized Feedback Control 121
- 7.1 Introduction 121
- 7.2 Pedestrian Flow Modeling 123
- 7.3 Feedback Linearization of State Equations 125
  - 7.3.1 Stability Under Feedback Linearizing Control 126
  - 7.3.2 Saturation of Control 128
- 7.4 Simulation Results 129
- 7.5 Code 133
- 7.6 Exercises 134
- 7.7 Computer Code 135
  - 7.7.1 main.m 135
  - 7.7.2 rhodot_nsec.m 138
  - 7.7.3 vfcntrl_nsec_try.m 138

8 Discretized Optimal Control 141
- 8.1 Optimal Control 142
  - 8.1.1 State Equations 142
  - 8.1.2 Cost Function 143
  - 8.1.3 Calculus of Variation 144

8.2		The Method of Steepest Descent	146
8.3		Numerical Results	147
8.4		Code	150
	8.4.1	main.m	150
	8.4.2	optimal_cntrl_calc_var_nsec_odesol.m	151
8.5		Exercises	152
8.6		Computer Code	152
	8.6.1	optimal_corridor_evacuation/main	152
	8.6.2	optimal_cntrl_calc_var_nsec_odesol	155

9 Distributed Feedback Control 1-D — 161

9.1		Introduction	162
9.2		Modeling	163
	9.2.1	One Equation Model	164
	9.2.2	Two Equation Model	165
9.3		Feedback Control for One-Equation Model	166
	9.3.1	Continuity Equation Control Model	166
	9.3.2	State Feedback Control	167
	9.3.3	Lyapunov Stability Analysis	168
	9.3.4	Simulation Results	170
9.4		Control Saturation	172
9.5		Feedback Control for Two Equation Model	175
	9.5.1	Two Equation Control Model	176
	9.5.2	State Feedback Control Using Backstepping	176
	9.5.3	Simulation	179
9.6		Exercises	182
9.7		Computer Code	182
	9.7.1	feedback_1d_mass	182
	9.7.2	feedback_1d_momentum	185

10 Distributed Feedback Control 2-D — 189

10.1		Introduction	189
10.2		Feedback Control of One-Equation Model	191
	10.2.1	One-Equation Model	191
	10.2.2	Control Model	192
	10.2.3	State Feedback Control	193
	10.2.4	Lyapunov Stability Analysis	193
	10.2.5	Simulation Results	195

Contents

    10.3 Feedback Control for Two-Equation Model 197
           10.3.1 Two Equation Model 197
    10.4 Control Model 198
           10.4.1 State Feedback Control Using Backstepping . 199
           10.4.2 Simulation Results 202
    10.5 Exercises 204
    10.6 Computer Code 204
           10.6.1 feedback_2d 204

11 Robust Feedback Control   **209**
    11.1 Introduction 209
    11.2 Feedback Control for Continuity Equation Model . . . 210
           11.2.1 Input Uncertain Control Model 211
           11.2.2 Robust Control by Lyapunov Redesign Method 212
           11.2.3 Simulation Results 216
    11.3 Robust Control for Two-Equation Model 220
           11.3.1 Robust Backstepping: Unmatched Uncertainty 221
           11.3.2 Robust Control: Matched Uncertainty 225
           11.3.3 Robust Control: Both Matched
                    and Unmatched Uncertainties 227
    11.4 Computer Code 229
           11.4.1 robust_1d 229

Bibliography   **233**

Index   **243**

Chapter 1

Introduction

The work presented in this book is concerned with developing pedestrian dynamic models that can be used in the control and the investigation of crowd behavior. The development of pedestrian dynamic models to implement evacuation system strategies is an ongoing research area. Since the early 1990s a strong interest in this topic has shown the importance of this issue [91]. Nevertheless, as stated in [103], our knowledge of the flow of crowds is inadequate and behind that of other transportation modes. Studies like the ones in [35, 92], are concerned with evacuation strategies for regional areas like cities and states. They are important in the decision making of an emergency evacuation such as in the case of a natural disaster. In smaller areas like airports, stadiums, theaters, buildings, and ships, an evacuation system is an important element in design safety [71]. A good evacuation system in the case of an emergency can prevent a catastrophic outcome as shown by [42].

It has been suggested in [73], that pedestrian traffic flow can be treated as vehicle traffic flow, where we can divide the problem into two categories: the microscopic level and the macroscopic level. The former involves individual units with characteristics such as individual speed and individual interaction. Microscopic pedestrian analysis studies were presented by [39, 46], and followed by many researchers to improve on the *Car-Following* model. In [44], the importance of a detailed design and pedestrian interactions is shown by implementing several case studies. On the other hand, the drawback to mathematical microscopic models is their difficult and expensive simulation, which leads to another approach called pedestrian analysis by simulation and is being increasingly used instead of the mathematical models [40]. In such models, pedestrians are treated as discrete individuals moving in a computer simulated environment. In this approach there are mainly three types of models, cellular based, physical force based, and queuing network based. More information on these microscopic models can be found in [10, 32, 38, 43, 45, 70, 78, 79, 94, 95, 101].

Macroscopic models aimed at studying pedestrian behavior use a continuum approach, where the movement of large crowds exhibit many of the attributes of fluid motion. As a result, pedestrian dynamics are treated as a fluid. This idea provides flexibility since detailed interactions are overlooked, and the model's characteristics are shifted toward parameters such as flow rate $f(\rho)$, concentration ρ (also known as traffic density), and average speed v, all being functions of 2-D space (x, y) and time (t). This class of models is classified under the hyperbolic partial differential equations. This way of modeling the pedestrian behavior was first introduced by [29, 30], and adopted by [11], where macroscopic models can be developed using (a) fluid flow theory, (b) a continuum responding to influences (local or non-local). The drawback to this way of modeling is due to the assumption that pedestrians behave similarly to fluids. Pedestrians tend to interact among themselves and with obstacles in their model area, which is not captured by macroscopic models.

To introduce feedback control as a strategy in evacuation plans, where the objective is to control each pedestrian in the model area is difficult and not practical for the microscopic modeling approach. The obvious reasons for the impracticality are the detailed design and pedestrian interactions which make it very difficult to control such models. Instead, the macroscopic modeling approach is well suited for understanding the rules governing the overall behavior of pedestrian flow for which individual differences are not that important. By monitoring panic situations like an evacuation, we see that (a) people tend to move as a group and, (b) behave in a rational manner, since they think the situation is fully visible to the tallest member of the group, who is supposed to convey information to the shorter members by his or her actions [51, 52].

1 Introduction

The first step in this study is to start investigating potential mathematical models governing crowd dynamics. Two factors need to be considered in deriving the models: the accurate representation of the pedestrian flow in an emergency situation (evacuation), and the complexity of the selected model.

In this book, four macroscopic crowd dynamic models have been developed. Their theoretical approaches are adopted to represent crowd motion in 2-D space (2-D). Four distinct 1-D traffic flow models are extended and modified to represent crowd dynamics in a 2-D environment. The resulting models are divided into two types, mainly one-equation and two equation crowd models (or systems) of nonlinear hyperbolic partial differential equations (PDE). These models start from conventional theory for ordinary fluids. All the models assume a conservation of continuity and in the last three, the second partial differential equation is a modified version of the momentum equation which differs in each model in order to distinguish them from fluid behavior.

The second step is to simulate the systems responses. For hyperbolic PDE's, computational fluid dynamics methods are used. These methods are aimed at finding the correct solution out of the many weak solutions available for the hyperbolic PDE. Here we applied first order accurate finite volume method schemes such as Lax-Friedrichs, FORCE, and Roe's schemes to find the models numerical solution. Then we designed a preliminary control to evacuate pedestrians based on feedback linearization applied directly on the PDE equation. We also discuss the issues and requirements of modeling, control, and communications for an intelligent evacuation system.

Chapter 2 will introduce the basic theory for the traffic flow problem in 1-D space along with the four macroscopic traffic flow models that will be used to develop the crowd models. In Chap. 3, new 2-D crowd models are presented, and the microscopic-to-macroscopic relationship for the last models is investigated, where a micro-to-macro link is established. Chapter 4 survey the state of the art numerical schemes based on computational fluid dynamic and gives the numerical solution of the models using Matlab software environment. Feedback linearization control for the 1-D space one-equation model is given in Chap. 5. In Chap. 6, we suggest an intelligent evacuation system, and discuss its control and information technology issues.

The following chapters show control design performed on discretized models of traffic as well as on the original distributed parameter models. Simulation studies and corresponding codes are provided to study the effectiveness of the control strategies.

Chapter 2

Traffic Flow Theory for 1-D

2.1 Introduction

Interest in modeling traffic flow has been around since the appearance of traffic jams. Ideally, if you can correctly predict the behavior of vehicle flow given an initial set of data, then, in theory, adjusting the flow in crucial areas can maximize the overall throughput of traffic along a stretch of road. This is of particular interest in regions of high traffic density, which may be caused by high volume peak time traffic, accidents or closure of one or more lanes of the road.

The development of the pedestrian evacuation dynamic systems follows from the traffic flow theory in 1-D space [2, 39, 51]. In many ways, the pedestrian evacuation system is similar to the vehicle traffic flow problem [73]. The main conservation equations used in modeling the vehicle traffic flow and the pedestrian evacuation flow are the same, with the exception that vehicle traffic is a 1-D space problem and the evacuation system is a 2-D space problem. Other similarities exist from having escape routes and escape times in both problems. In most of the situations a vehicle or a pedestrian has more than one route to a destination and each route has an associated cost, such as time.

In this chapter we will give the necessary background on traffic flow theory and survey the existing macroscopic mathematical models for single-lane, 1-D space traffic flow. These models will be used for crowd flow in 1-D, and they will be modified in Chap. 3 for 2-D flow. In Sect. 2.2, we start with the concept of macroscopic vs microscopic ways of modeling the traffic flow problem, followed by Sect. 2.3, where a microscopic model is introduced. The derivation of the traffic flow theory based on conservation of mass law, and the relationships between velocity and density are given in Sect. 2.4. In Sect. 2.5, four macroscopic traffic flow models are presented, derived, and analyzed based on their mathematical characteristics. Finally, the exact and weak solutions to the scalar traffic flow PDE, and the concepts of shock wave, rarefaction wave, and the admissibility of a solution are considered.

2.2 Microscopic vs Macroscopic

In the traffic flow problem, there are two classes of models: Macroscopic, which is concerned with average behavior, such as traffic density, average speed and module area, and a second class of models based on individual behavior referred to as microscopic models. The latter is classified into different types. The most famous one is the *Car-Following* models [6, 17, 57], where the driver adjusts his or her acceleration according to the conditions in front. In these models the vehicle position is treated as a continuous function and each vehicle is governed by an ordinary differential equation (ODE) that depends on speed and distance of the car in front. Another type of microscopic models are the *Cellular Automata* or vehicle hopping which differs from *Car-Following* in that it is a fully discrete model. It considers the road as a string of cells which are either empty or occupied by one vehicle. One such model is the *Stochastic Traffic Cellular Automata*, given in [75]. However, microscopic approaches are computationally expensive, as each car has an ODE to be solved at each time step, and as the number of cars increases, so does the size of the system to be solved. On the other hand, the macroscopic models are computationally less expensive because they have fewer design details in terms of interaction among vehicles and between vehicles and their environment. Therefore, it is desirable to use macroscopic models if a good model can be found satisfactorily to describe the traffic flow. In addition, this idea provides flexibility since detailed interactions are overlooked, and the model's characteristics are shifted toward parameters such as flow rate $f(\rho, v)$, concentration ρ (also known as traffic density), and average speed v, all functions of 1-D or 2-D space (x, y), and time (t). This is also true for first-order fluid dynamic models of isothermal flow and gases through pipes.

Two main prototypes set the stage for macroscopic traffic flow: the first is called the LWR model which is a *non-linear*, first-order hyperbolic PDE based on law of conservation of mass. The second one is a *second-order* model known as the PW model, which is based on two coupled PDE's one given by the conservation of mass and a second equation that mimics traffic flow.

2.3 Car-Following Model

We present the well known *car-following* microscopic traffic flow model. In [93], a 2-D version of this model was used for pedestrian flow in 2-D space. To derive the 1-D model, first assume cars can not pass each other. Then the idea is that a car in 1-D can move and accelerate forward based on two parameters; the headway distance between the current car and the one in front, and their speed difference. Hence, it is called *following*, where a car from behind follows the one in front, and this is the anisotropic property. This property is also desirable in macroscopic models, since it reflects the actual observed behavior of traffic flow [23].

Suppose the nth car location is $x_n(t)$, then the nonlinear model is given by

$$\ddot{x}_n(t) = c \frac{\dot{x}_n(t) - \dot{x}_{n-1}(t)}{x_n(t) - x_{n-1}(t)}, \qquad (2.1)$$

The acceleration of the current car $\ddot{x}_n(t)$ depends on the front car speed and location, c is the sensitivity parameter. Integrating the above yields

$$\dot{x}_n(t) = c \ln\left(x_n(t) - x_{n-1}(t)\right) + d_n. \qquad (2.2)$$

Since by the definition of the density (number of cars per unit area)

$$\frac{1}{\rho(x,t)} = x_n(t) - x_{n-1}(t), \qquad (2.3)$$

and the integration constant d_n is chosen such that at jam density ρ_m, the velocity is zero. Then for steady-state we get

$$v = -c \ln \frac{\rho}{\rho_m}. \qquad (2.4)$$

We see that for $\rho \to 0$ we get in trouble, but from observations in low traffic densities, car speed is the maximum allowed speed, hence we can assume $v = v_{\max}$, which is the maximum allowed speed.

2.4 Traffic Flow Theory

In this section we will cover the vehicle traffic flow fundamentals for the macroscopic modeling approach. The relation between density, velocity and flow is presented for traffic flow. Then we derive the conservation of vehicles, which is the main governing equation for scalar macroscopic traffic models. Finally, the velocity–density functions that makes the conservation equation a function of only one variable (density) are given.

2.4.1 Flow

In this section, we will illustrate the close relationship between the three variables: density, velocity and traffic flow. Suppose there is a road with cars moving with constant velocity v_0, and constant density ρ_0 such that the distance between the cars is also constant as shown in the Fig. 2.1a. Now let an observer measure the number of cars per unit time τ that pass him (i.e. traffic flow f). In τ time, each car has moved $v_0 \tau$ distance, and hence the number of cars that pass the observer in τ time is the number of cars in $v_0 \tau$ distance, see Fig. 2.1b.

Since the density ρ_0 is the number of cars per unit area and there is $v_0 \tau$ distance, then the traffic flow is given by

$$f = \rho_0 v_0 \qquad (2.5)$$

This is the same equation as in the time varying case, i.e.,

$$f(\rho, v) = \rho(x, t) v(x, t). \qquad (2.6)$$

2.4 Traffic Flow Theory

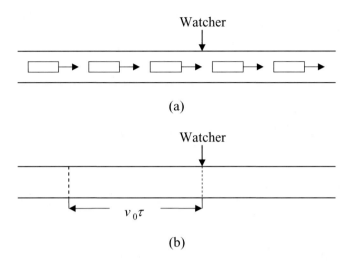

Fig. 2.1. (a) Constant flow of cars; (b) Distance traveled in τ hours for a single car

To show this, consider the number of cars that pass point $x = x_0$ in a very small time Δt. In this period of time the cars have not moved far and hence $v(x,t)$, and $\rho(x,t)$ can be approximated by their constant values at $x = x_0$ and $t = t_0$. Then, the number of cars passing the observer occupy a short distance, and they are approximately equal to $\rho(x,t)v(x,t)\Delta t$, where the traffic flow is given by (2.6).

2.4.2 Conservation Law

The models for traffic, whether they are one-equation or system of equations, are based on the physical principle of *conservation*. When physical quantities remain the same during some process, these quantities are said to be conserved. Putting this principle into a mathematical representation will make it possible to predict the densities and velocities patterns at future time. In our case, the number of cars in a segment of a highway $[x_1, x_2]$ are our physical quantities, and the process is to keep them fixed (i.e., the number of cars coming in equals the number of cars going out of the segment). The derivation of the conservation law is given in [26, 37], and it is presented here for completion. Consider a stretch of highway on which cars are

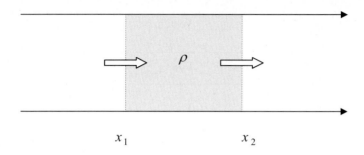

Fig. 2.2. One-dimension flow

moving from left to right as show in Fig. 2.2. It is assumed here that there are no exit or entrance ramps. The number of cars within $[x_1, x_2]$ at a given time t is the integral of the traffic density given by

$$N = \int_{x_1}^{x_2} \rho(x,t)\,dx. \tag{2.7}$$

In the above equation, it is implied that the number of people within $[x_1, x_2]$ is at maximum when traffic density is equal to jam density ρ_m which is associated with the maximum number of cars that could possibly fit in a unit area.

The number of cars can still change (increase or decrease) in time due to cars crossing both ends of the segment. Assuming no cars are crated or destroyed, then the change of the number of cars is due to the change at the boundaries only. Therefore, the rate of change of the number of cars is given by

$$\frac{dN}{dt} = f_{\text{in}}(\rho, v) - f_{\text{out}}(\rho, v), \tag{2.8}$$

since the number of cars per unit time is the flow $f(\rho, v)$. Combining (2.7), and (2.8), yields the **integral conservation law**

$$\frac{d}{dt}\int_{x_1}^{x_2} \rho(x,t)\,dx = f_{\text{in}}(\rho, v) - f_{\text{out}}(\rho, v). \tag{2.9}$$

This equation represents the fact that change in number of cars is due to the flows at the boundaries. Now let the end points be independent variables (not fixed with time), then the full derivative is replaced

2.4 Traffic Flow Theory

by partial derivative to get

$$\frac{\partial}{\partial t}\int_{x_1}^{x_2} \rho(x,t)\,dx = f_{\text{in}}(\rho,v) - f_{\text{out}}(\rho,v). \tag{2.10}$$

The change in the number of cars with respect to distance is given by

$$f_{\text{in}}(\rho,v) - f_{\text{out}}(\rho,v) = -\int_{x_1}^{x_2} \frac{\partial f}{\partial x}(\rho,v)\,dx, \tag{2.11}$$

and by setting the last two equations equal to each other, we get

$$\int_{x_1}^{x_2} \left[\frac{\partial \rho}{\partial t}(x,t) + \frac{\partial f}{\partial x}(\rho,v)\right] dx = 0. \tag{2.12}$$

This equation states that the definite integral of some quantity is always zero for all values of the independent varying limits of the integral. The only function with this feature is the zero function. Therefore, assuming $\rho(x,t)$, and $q(x,t)$ are both smooth, the **1-D conservation law** is found to be

$$\rho_t + f_x(\rho,v) = 0. \tag{2.13}$$

We need to mention that this equation is valid for traffic and many more physical quantities. The idea here is conservation, and for vehicle traffic flow, the flow is given by (2.6).

2.4.3 Velocity–Density Relationship(s)

Traffic density and vehicle velocity are related by one equation, conservation of vehicles,

$$\rho_t(x,t) + (\rho(x,t)v(x,t))_x = 0, \tag{2.14}$$

where the notation $(\cdot)_\phi = \frac{\partial(\cdot)}{\partial \phi}$ will be used from here on. If the initial density and the velocity field are known, the above equation can be used to predict future traffic density. This leads us to choose the velocity function for the traffic flow model to be dependent on density and call it $V(\rho)$. The choice of such function depends on the behavior the model is trying to mimic. The following is a brief description of models that have been recognized and used by researchers [57], with emphasis on Greenshield model that will be used in several traffic (crowd) models throughout this book.

Greenshield's Model [34]

This model is simple and widely used. It is assumed here that the velocity is a linearly decreasing function of the traffic flow density, and it is given by

$$V(\rho) = v_f(1 - \frac{\rho}{\rho_m}) \qquad (2.15)$$

where v_f is the free flow speed and ρ_m is the maximum density. Figure 2.3 shows the speed $V(\rho(x,t))$ as a monotonically decreasing function. For zero density the model allows free flow speed v_f, while for maximum density ρ_m no car can move in or out.

The flux–density relationship for Greenshield's model (2.15) is given in Fig. 2.4, where it shows the flux increases to a maximum which occurs at some density $\hat{\rho}$ and then it goes back to zero. This kind of behavior is due to the fact that $f''(\rho) < 0$ (note that $f(\rho) = \rho V(\rho)$ is the flux flow).

Greenberg Model [33]

In this model the speed–density function is given by

$$V(\rho) = v_f \ln(\frac{\rho}{\rho_m}) \qquad (2.16)$$

Underwood Model [33]

In the Underwood model the velocity–density function is represented by

$$V(\rho) = v_f \exp(\frac{-\rho}{\rho_m}) \qquad (2.17)$$

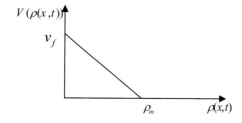

Fig. 2.3. Greenshield's model for traffic flow speed

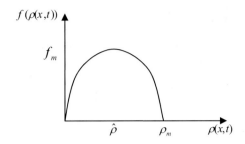

Fig. 2.4. Traffic flow flux as a function of density

Diffusion Model [14, 74]

Diffusion is a good extension to the model given by (2.15), where the effect of gradual rather than instantaneous reduction of speed by the driver takes place in response to shock waves. This kind of reaction can be accomplished by adding an extra term such that the modified Greenshield model will become

$$V(\rho) = v_f(1 - \frac{\rho}{\rho_m}) - \frac{D}{\rho}(\frac{\partial \rho}{\partial x}) \qquad (2.18)$$

where D is a diffusion coefficient given by

$$D = \tau \, v_r^2$$

and v_r is a random velocity, τ is a relaxation parameter.

2.5 Traffic Flow Model 1-D

In this section, we will present four different models for traffic flow in 1-D space. The first model is one equation model, and the rest are systems of two-equation models. All the models are described by partial differential equations and based on conservation of mass and a second equation that is intended to capture the complex interactions observed in traffic flow motion. In addition, this second equation provides another way to couple velocity and density. Hence the flow is not in equilibrium like the one equation model. This is one of the main differences between the scalar and the system models.

2.5.1 LWR Model

The first model used in describing the traffic flow problem known as the LWR model, named after the authors in [68] and [87]. The LWR model is a scalar, time-varying, non-linear, hyperbolic partial differential equation. The model governing equation is (2.14). In this equation, traffic density is the conserved quantity, and we rewrite the model as

$$\rho_t(x,t) + (\rho(x,t)V(\rho(x,t)))_x = 0 \qquad (2.19)$$

with the flux being replaced by the velocity–density relationship

$$f(x,t) = \rho(x,t)V(\rho(x,t)) \qquad (2.20)$$

and $V(\rho(x,t))$ is the velocity function given by (2.15).

One of the basic assumptions in the LWR model regarding the velocity is its dependence on density alone. Any changes to density will be reflected in the velocity. The drawback for this assumption, as pointed out in [23], is that traffic is in equilibrium when such velocity–density functions are used, i.e., given a particular density, especially for light traffic, the velocity will be fixed and the model does not recognize that there is a distribution of desired velocities across vehicles. Therefore the model is not able to describe observed behavior in light traffic, although one can argue that v_f is an average speed which might take care of this issue. On the plus side, the model is anisotropic as the nature of the observed traffic flow, i.e. vehicle behavior is affected by mostly the car in front. This can be found from the model eigenvalue given by

$$f'(\rho(x,t)) = \frac{\partial f}{\partial \rho}(\rho(x,t)) = V(\rho(x,t)) + \rho V'(\rho(x,t)), \qquad (2.21)$$

which means that the model allows information to travel as fast as the flow of traffic, and not more, since it satisfies $0 < f'(\rho) < V(\rho)$, because

$$V'(\rho) = -\frac{v_f}{\rho_m}. \qquad (2.22)$$

The LWR model given by (2.19) and (2.15) is a simple model and it is unable to capture all of the complex interactions for a realistic traffic flow model. For this reason, modifications to the LWR model have been suggested. One way is by the various velocity–density

2.5 Traffic Flow Model 1-D

functions we gave in Sect. 2.4.3. The second way is by coupling the conservation of mass with a second equation that tries to mimic traffic motion instead of the velocity–density models as given next.

2.5.2 PW Model

The first system model to be presented is a two-equation model proposed in the 1970s independently in [82] and [102]. Their model was the first model to couple velocity dynamics as a second equation, and it is referred to as the PW model. The first equation is the conservation of mass as discussed in the previous sections

$$\rho_t + (\rho v)_x = 0, \qquad (2.23)$$

where the flux function $f(\rho, v) = \rho v$. In the LWR scalar equation model, a particular form of v was assumed where velocity is a function of density, but in high order models, v and ρ are assumed to be independent and a second equation is formed to link them, as in fluid and gas models. The second equation is derived from the Navier-Stokes equation of motion for a 1-D compressible flow, but with the pressure term replaced by $P = C_0^2 \rho$, where C_0 is the anticipation term that describes the response of macroscopic driver to traffic density, i.e. space concentration, and the pressure now is not "pressure" as such. The model also includes a traffic relaxation term that keeps speed concentration in equilibrium

$$\frac{V(\rho) - v}{\tau},$$

where τ is a relaxation time, and the velocity $V(\rho)$ is the maximum out-of-danger velocity meant to mimic driver's behavior given earlier by (2.15) to (2.18). The second equation of the PW model in nonconservative form is then given by

$$v_t + v\, v_x = \frac{V(\rho) - v}{\tau} - \frac{C_0^2}{\rho} \rho_x. \qquad (2.24)$$

To study this model, we have to find its eigenvalues by first rewriting the model in conservation form. The first step is to use the product rule

$$(\rho v)_t = \rho v_t + v \rho_t, \qquad (2.25)$$

and by multiplying (2.23) by v, we get

$$v\rho_t + v(\rho v)_x = 0. \qquad (2.26)$$

Then by substituting in for $v\rho_t$ from the product rule (2.25), we get

$$v(\rho v)_x + (\rho v)_t - \rho v_t = 0, \qquad (2.27)$$

and by substituting (2.27) into (2.24), and multiply the result by ρ we get

$$\rho v_t + \rho v\, v_x = \rho\left(\frac{V(\rho) - v}{\tau}\right) - C_0^2 \rho_x.$$

Then by substituting for ρv_t from (2.27) we get

$$v(\rho v)_x + (\rho v)_t + \rho v\, v_x = \rho\left(\frac{V(\rho) - v}{\tau}\right) - C_0^2 \rho_x. \qquad (2.28)$$

Again using the product rule on $(\rho v v)_x$, i.e.,

$$(\rho v v)_x = (\rho v)_x v + (\rho v) v_x \qquad (2.29)$$

and substituting in (2.28) we obtain

$$(\rho v^2)(\rho v)_t + (\rho v^2)_x = \rho\left(\frac{V(\rho) - v}{\tau}\right) - C_0^2 \rho_x. \qquad (2.30)$$

Hence we obtain (2.24) with the lefthand side in conservation form

$$(\rho v)_t + (\rho v^2 + C_0^2 \rho)_x = \rho\left(\frac{V(\rho) - v}{\tau}\right) \qquad (2.31)$$

where now ρ and ρv are the conserved variables. Equations (2.23) and (2.31) can be written in vector form as

$$Q_t + F(Q)_x = S \qquad (2.32)$$

where,

$$Q = \begin{bmatrix} \rho \\ \rho v \end{bmatrix},\ F(Q) = \begin{bmatrix} \rho v \\ \rho v^2 + C_0^2 \rho \end{bmatrix},\ S = \begin{bmatrix} 0 \\ \rho\left(\dfrac{V(\rho) - v}{\tau}\right) \end{bmatrix}. \qquad (2.33)$$

2.5 Traffic Flow Model 1-D

Setting the source term $S = 0$, we can rewrite the system in quasi-linear form as

$$Q_t + A(Q)\, Q_x = 0, \qquad (2.34)$$

where

$$A(Q) = \frac{\partial F}{\partial Q} = \begin{bmatrix} 0 & 1 \\ C_0^2 - \rho v^2 & 2v \end{bmatrix}. \qquad (2.35)$$

Finally by solving for the eigenvalues from

$$|A(Q) - \lambda I| = 0 \qquad (2.36)$$

we get two distinct and real eigenvalues

$$\lambda_{1,2} = v \pm C_0, \qquad (2.37)$$

therefore, the system is strictly hyperbolic.

The model has a major drawback that researchers (see for example [23]) are concerned about, mainly, that the model strongly follows the fluid flow theory. In fluids, the behavior of a particle is affected by its surrounding particles. Thus the anisotropic nature of traffic is not preserved since the vehicles are allowed to move with negative velocity, i.e. against the flow. This is clear from the eigenvalue, where one of $\lambda_{1,2} = v \pm C_0$ is always greater than the vehicle speed v. So, information from behind affects the behavior of the driver, and this is not true for observed traffic flow. This is called the isotropic property.

2.5.3 AR Model

A new model in [4] and improved in [84] is argued to be an improvement on the PW model. The authors of this model say that other researchers have stuck too closely to fluid flow models and have not allowed for a significant difference between traffic and fluids, e.g., traffic is more concerned with the flow in front, rather than behind. Therefore in order to move away from fluids and toward the anisotropic property of traffic, they argue that replacing the "pressure" term with an anticipation term describing how the average driver behaves is not a sufficient fix for the differences between the two types of flow. They claim that the drawback in the PW model (letting information travel faster than the flow) is due to an incorrect anticipation factor involving the derivative of the pressure w.r.t. x. Therefore they suggest the correct dependence must involve the convective derivative (full derivative) of the pressure term. The convective derivative in its general form is given by

$$\frac{D\phi}{Dt} = \frac{\partial \phi}{\partial t} + (\vec{v} \cdot \nabla)\phi, \qquad (2.38)$$

for $\phi(\vec{x}, t)$, and $\vec{x} \in \Re^n$. They support their claim by the following example: "Assuming that in front of a driver traveling with speed v the density is increasing with respect to x, but decreasing with respect to $(x-vt)$. Then the PW type models predict that this driver would slow down, since the density ahead is increasing with respect to x! On the contrary, any reasonable driver would accelerate, since this denser traffic travels faster than him."

We call the model AR for short, and the first PDE equation is the same conservation of cars given by (2.23). However, in the AR model the next lagrangian equation replaces the second PDE equation given in the PW model. This second equation is found by applying the full derivative (2.38) to describe traffic motion dynamics, and it is given by

$$(v + P(\rho))_t + v\,(v + P(\rho))_x = 0, \qquad (2.39)$$

where $P(\rho)$ is an increasing function of density. This choice is to ensure that this model caries the anisotropic property, and it is given by

$$P(\rho) = C_0^2\, \rho^\gamma, \qquad (2.40)$$

where $\gamma > 0$, and $C_0 = 1$. Next we will put the second equation in conservation form, and find the system eigenvalues. First multiply (2.39) by ρ, then by using the product rule

$$(\rho\,(v + P(\rho)))_t = \rho_t(v + P(\rho)) + \rho\,(v + P(\rho))_t, \qquad (2.41)$$
$$(\rho v\,(v + P(\rho)))_x = (\rho v)_x(v + P(\rho)) + (\rho v)(v + P(\rho))_x, \qquad (2.42)$$

we obtain

$$(\rho\,(v+P(\rho)))_t - \rho_t(v+P(\rho)) + (\rho v\,(v+P(\rho)))_x - (\rho v)_x(v+P(\rho)) = 0. \qquad (2.43)$$

Now, using the conservation law (2.23) for ρ_t, we can simplify the above equation to

$$(\rho\,(v + P(\rho)))_t + (\rho v\,(v + P(\rho)))_x = 0, \qquad (2.44)$$

2.5 Traffic Flow Model 1-D

which is the conservation form of (2.39). Then our conserved variables are ρ, and $\rho(v + P(\rho))$. We proceed now to find the system eigenvalues, let $X = \rho(v + P(\rho))$ for simplification, then the AR model given by (2.23), and (2.44) can be rewritten as

$$\begin{cases} \rho_t + (X - \rho P(\rho))_x = 0 \\ X_t + \left(\dfrac{X^2}{\rho} - X P(\rho)\right)_x = 0 \end{cases} \quad (2.45)$$

and in vector form (2.32), the stats and the flux are given by

$$Q = \begin{bmatrix} \rho \\ X \end{bmatrix}, \quad F(Q) = \begin{bmatrix} X - \rho P(\rho) \\ \dfrac{X^2}{\rho} - X P(\rho) \end{bmatrix}. \quad (2.46)$$

For the quasi-linear form (2.34), the Jacobian is given by

$$A(Q) = \frac{\partial F}{\partial Q} = \begin{bmatrix} -(\gamma+1) & 1 \\ -\left(\dfrac{X^2}{\rho^2} + \dfrac{\gamma X P(\rho)}{\rho}\right) & \left(\dfrac{2X}{\rho} - P(\rho)\right) \end{bmatrix}. \quad (2.47)$$

Finally, solving for the eigenvalues from $|A(Q) - \lambda I| = 0$, we find two distinct and real eigenvalues

$$\lambda_1 = v - \gamma P(\rho) \quad \& \quad \lambda_2 = v. \quad (2.48)$$

Therefore, the system is strictly hyperbolic and since the "pressure" is an increasing function, then it is guaranteed that $\lambda_1 < \lambda_2$ due to the fact that the maximum wave speed is equal to the velocity of the flow v. Hence, the anisotropic property of traffic is preserved.

2.5.4 Zhang Model

We present here a model that was proposed in [104, 105], and it claims not to be of fluid, or gas-like behavior. The model caries the anisotropic property, because the second equation is derived from the microscopic *car-following* model. Hence, a micro-to-macro link is established for this model. Again, as in the PW and AR models, the Zhang model is also a system consisting of the conservation of cars (2.23), and coupled with a second PDE that describe car motion given by

$$v_t + v v_x + \rho V'(\rho) v_x = 0 \quad (2.49)$$

We start the micro-to-macro derivation from the *homogeneous* microscopic *car-following* model (the relaxation term can be added for 2-D crowd flow given in the next chapter) given by

$$\tau(s_n(t))\ddot{x}_n(t) = \dot{x}_{n-1}(t) - \dot{x}_n(t), \tag{2.50}$$

where

$$s_n(t) = x_{n-1}(t) - x_n(t), \tag{2.51}$$

and $s_n(t)$ is a function of the local spacing between cars, $x_n(t)$ is the position of the nth car, $\ddot{x}_n(t)$ is the acceleration, $\dot{x}_n(t)$ is the velocity, and $\tau(s_n(t))$ is the average response time to the headway distance. Using the above notations, we rewrite (2.50), and define the velocity as $v(x,t) = \dot{x}(t)$ to obtain the following

$$\tau(s(x(t),t))\frac{dv(x,t)}{dt} = \frac{d(s(x(t),t))}{dt}, \tag{2.52}$$

and by using convective derivative $\partial_t + v\partial_x$ on the velocity component, we get

$$\tau(s)(v_t + vv_x) = (s_t + vs_x). \tag{2.53}$$

From the conservation law (2.23), let $\rho = 1/s$, and by using the following derivative form

$$D_x\left(\frac{a}{b}\right) = \frac{bD_xa - aD_xb}{b^2}, \tag{2.54}$$

for any a and $b \neq 0$, we get

$$s_t + vs_x + us_y = sv_x + su_y, \tag{2.55}$$

and by direct substituting in the right hand side of (2.53), we obtain our desired equation in the following form

$$(v_t + vv_x) = \frac{s}{\tau(s)}v_x, \tag{2.56}$$

where

$$\frac{s}{\tau(s)} = -C(\rho) = -\rho V'(\rho) \geq 0 \tag{2.57}$$

is the sound wave speed. This completes the derivation of the macroscopic model (2.49) from its microscopic counterpart.

2.5 Traffic Flow Model 1-D

The conservative form of this model is derived next. First collect terms and rewrite (2.49) to get

$$v_t + (v + \rho V'(\rho)) v_x = 0, \tag{2.58}$$

then expand the conservation of mass equation

$$\rho_t + \rho v_x + v \rho_x = 0. \tag{2.59}$$

We substitute for ρv_x from (2.59) into (2.58) to obtain

$$v_t + v v_x + V'(\rho)(-\rho_t - v\rho_x) = 0, \tag{2.60}$$

which can be rewritten as

$$v_t + v v_x - (V(\rho))_t - v(V(\rho))_x = 0, \tag{2.61}$$

or in lagrangian form as

$$(v - V(\rho))_t + v(v - V(\rho))_x = 0. \tag{2.62}$$

We now proceed to find the conservation form by multiplying (2.62) by ρ and using the product rules

$$\begin{aligned}(\rho(v - V(\rho)))_t &= \rho_t(v - V(\rho)) + \rho(v - V(\rho))_t, & (2.63)\\ (\rho v(v - V(\rho)))_x &= (\rho v)_x(v - V(\rho)) + \rho v(v - V(\rho))_x, & (2.64)\end{aligned}$$

to get

$$(\rho(v-V(\rho)))_t - \rho_t(v-V(\rho)) + (\rho v(v-V(\rho)))_x - (v\rho)_x(v-V(\rho)) = 0. \tag{2.65}$$

From (2.59), we substitute for ρ_t in the above equation to obtain our final conservation form given by

$$(\rho(v - V(\rho)))_t + (\rho v(v - V(\rho)))_x = 0, \tag{2.66}$$

where our states are given by ρ and $\rho(v-V(\rho))$. For the quasi-linear form (2.34), let $X = \rho(v - V(\rho))$, and write the system in vector form (2.32), such that

$$Q = \begin{bmatrix} \rho \\ X \end{bmatrix}, \quad F(Q) = \begin{bmatrix} X + \rho V(\rho) \\ \dfrac{X^2}{\rho} - XV(\rho) \end{bmatrix}. \tag{2.67}$$

The Jacobian is then can be found to be

$$A(Q) = \frac{\partial F}{\partial Q} = \begin{bmatrix} V(\rho) + \rho V'(\rho) & 1 \\ -\left(\frac{X^2}{\rho^2} - XV'(\rho)\right) & \left(\frac{2X}{\rho} + V(\rho)\right) \end{bmatrix}. \quad (2.68)$$

Finally by solving for the eigenvalues from $|A(Q) - \lambda I| = 0$, we find two distinct and real eigenvalues

$$\lambda_1 = v + \rho V'(\rho) \quad \& \quad \lambda_2 = v \quad (2.69)$$

therefore the system is strictly hyperbolic. Since $V'(\rho)$ is negative and given by (2.22), then the maximum the information can travel is equal to the vehicle speed v.

2.5.5 Models Summary

The one-equation LWR model consists of a single wave whose velocity is given by the derivative of the flux function, and information travels forward at a maximum not faster than the speed of traffic. Therefore the model behavior is anisotropic, i.e. only reacts to conditions ahead. For the two-equation models, the PW has two waves traveling at speeds given by $v \pm C_0$, one of them will always be traveling faster than the current speed v. This is a major cause of criticism of this model. The AR model has wave speeds given by v and $v - \gamma P(\rho)$. This seems reasonable since, as in the LWR model, the faster wave will move at the same speed as the traffic v. This is also true for the Zhang model, whose wave speeds are given by v and $v + \rho V'(\rho)$ with $V'(\rho) \leq 0$. This demonstrates the desirable anisotropic nature of the LWR, AR and Zhang models, and the isotropic nature of the PW model, for which it is severely criticized. In addition, Zhang model has a microscopic counterpart, which is not true for the PW and AR models. Although, the AR model has an indirect micro-to-macro relationship [4], where a numerical discretization of the AR model and a microscopic *car-following* model gave the same numerical formula. This suggest that the macroscopic AR model can be considered as an upper limit to the microscopic model.

2.6 Method of Characteristics

A typical problem in partial differential equation consists of finding the solution of a PDE subject to boundary conditions (BVP), initial conditions (IVP), or both (IBVP). In most cases it is difficult to find the exact (classical) solution of a hyperbolic PDE, but due to the simplicity of the LWR model and the fact it is a scalar 1-D space model we are able to find the exact solution by method of characteristics. The method of characteristics is a widely used technique to solve hyperbolic PDE's [24, 58, 77, 86].

2.6.1 LWR Model Classification

The partial derivative scalar conservation law in (2.19) is classified as first-order quasi-linear partial differential equation. This is due to the fact that the derivative of the highest partial occurs linearly. We can rewrite (2.19) as

$$\rho_t(x,t) + f'(\rho(x,t))\,\rho_x = 0 \qquad (2.70)$$

where $f'(\rho)$ is the vehicle speed, and it is called the characteristic slop or the eigenvalue of the PDE. By using Greenshield's model (2.15), we get

$$f'(\rho(x,t)) = \frac{\mathrm{d}f(\rho(x,t))}{\mathrm{d}x} = v_f - \frac{2\,v_f\,\rho(x,t)}{\rho_m}, \qquad (2.71)$$

where we see that this eigenvalue is real. Therefore, the LWR model is classified as strictly hyperbolic PDE. Figure 2.5 below shows the changes in speed $f'(\rho)$ with respect to the changes in density ρ. This relationship is important in finding the solution to the traffic flow model (2.70) by using method of characteristics discussed in the following section.

2.6.2 Exact Solution

Here we will use the method of characteristics to solve the initial value problem (IVP) and also called the Cauchy problem given by

$$\begin{cases} \rho_t(x,t) + f(\rho(x,t))_x = 0 \\ \rho(x,0) = \rho_0(x) \end{cases} \qquad (2.72)$$

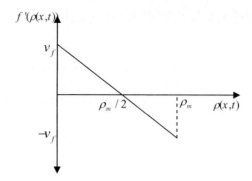

Fig. 2.5. Characteristic slops vs density

where $x \in \Re$ and time $t \in \Re^+$. Flow rate $f : \Re \mapsto \Re$ is assumed to be a smooth function, at least C^2 (i.e., twice differentiable) and the initial condition $\rho_0 : \Re \mapsto \Re$ is continuous. For the single conservation law, the eigenvalue of the PDE in (2.72) is given by the slope of the characteristic curve found from the quasi-linear form (2.70) as

$$\lambda(\rho) = f'(\rho) \tag{2.73}$$

Theorem 2.6.1 *Any C^1 solution of the single conservation law in (2.72) is constant along its characteristics. Accordingly, characteristic curves for the partial derivative conservation law in (2.72) are straight lines.*

Proof See [86].

The above theorem implies that any curve of the form

$$x(t) = k\,t + x(0) \tag{2.74}$$

is a characteristic curve where $x(t)$ is the solution, $k = f'(\rho(x(t), t))$ is the constant slope of the characteristic rays and $x(0)$ is the initial position of the characteristics rays. To show that (2.74) is indeed a solution to our Cauchy problem, let us first define what is meant by a solution.

Definition 2.6.1 Let $f : \Re \mapsto \Re$ be smooth, and let $\rho_0 : \Re \mapsto \Re$ be continuous. We say that $\rho(x,t) : (\Re \times \Re^+) \mapsto \Re$ is a classical solution of the Cauchy problem if $\rho(x,t) \in C^1(\Re \times \Re^+) \cap C^0(\Re \times \Re^+)$ and (2.70) is satisfied.

2.6 Method of Characteristics

We can verify that it is indeed a solution by substituting for $x(0) = x - f't$ from (2.74) to get

$$\rho(x,t) = \rho_0(x - f't), \tag{2.75}$$

then, by taking partial derivatives with respect to t and x, respectively, we obtain

$$\rho_t = \rho_0'(x - f't)(-f'), \quad \text{and} \quad \rho_x = \rho_0'(x - f't).$$

Substituting the above in (2.70), we get

$$\rho_0'(x - f't)(-f') + f'\rho_0'(x - f't) = 0,$$

which shows (2.72) is satisfied. So, the exact solution is basically the initial data shifted by the slope of the characteristic as shown in Fig. 2.6.

From the initial data we are able to generate the slopes of the characteristic rays originating from the x-axis. For some certain profiles of initial data, this gives us a method for solving the Cauchy problem. To illustrate this method, let us consider the initial data in Fig. 2.7, where we have the density profile of heavy traffic density at one end and light at the other end. Substituting for the initial density values in (2.71) will give the slopes of each of the characteristics. Then the solution follows from (2.74) along the rays of the characteristics as shown on the same figure.

2.6.3 Blowup of Smooth Solutions

Unfortunately, other examples with different initial data show how easily the procedure above fails. In Fig. 2.8, we see a density profile

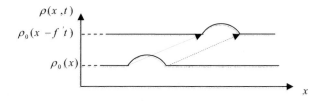

Fig. 2.6. Exact solution is achieved by shifting the initial density profile

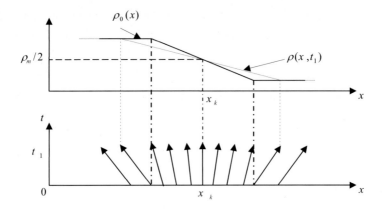

Fig. 2.7. Density distribution in the upper part, and the corresponding characteristic rays in the lower part

that describes road condition when cars are approaching red traffic light. Although initial density is continuous, the characteristics overlap at some later point in time. Since our solution cannot be multi-valued, we must conclude (in light of Theorem 2.6.1) that the solution shown cannot be smooth. For this type of initial data, a theory of discontinuous solutions, or *shock wave* solution is used.

Moreover, in Fig. 2.9, we face a different kind of problem. This time we have initial profile corresponding to heavy traffic at the beginning, then at some point x_k it is lighter. This example can

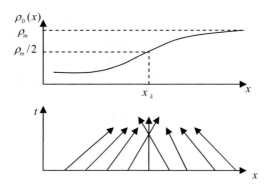

Fig. 2.8. Overlapping characteristics from continuous initial data

2.6 Method of Characteristics

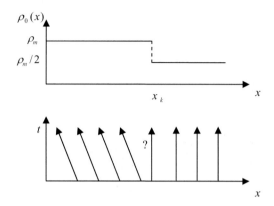

Fig. 2.9. Characteristics do not specify solution in the wedge

be related to conditions of a red light turns to green. As we all observe in real situation, cars start to accelerate from high density (low speed) to low density (higher speed). The exact solution for this data shows that there is a region untouched by any characteristics from the given initial data. Thus, the method of characteristics did not identify a solution in this region. As we shall see next, for this case we will be able to identify a continuous solution called a *rarefaction or fan wave* solution to fill the wedge.

2.6.4 Weak Solution

From the discussion above, smooth solutions of a single conservation laws can blow up (develop discontinuities or singularities) in finite time which fails to make any sense. Therefore, one cannot follow the practice of accepting solutions to the hyperbolic partial differential equations as given directly by method of characteristics. In order to understand discontinuous solutions, one needs to extend the notion of solution itself. One of the main features of the quasi-linear theory for hyperbolic PDE's is the notion of *weak solutions*. For a given initial data,

Definition 2.6.2 Let $\rho_0 \in L^\infty$. Then ρ is a weak solution or a solution in the distributional sense of (2.72) if and only if $\rho \in L^\infty(\Re \times \Re^+)$ and,

$$\int_0^\infty \int_{-\infty}^\infty [\rho(x,t)\,\phi_t(x,t) + f(\rho(x,t))\,\phi_x(x,t)]\,\mathrm{d}x\,\mathrm{d}t$$
$$+ \int_{-\infty}^\infty \rho_0(x)\,\phi(x,0)\,\mathrm{d}x = 0 \quad (2.76)$$

is satisfied for every $\phi \in C_0^\infty(\Re \times [0,\infty[)$
and $C_0^\infty(\Re \times [0,\infty[) := \{C_0^\infty(\Re \times [0,\infty[) \mid \exists\, r > 0 \text{ s.t. support of } \phi \subset B_r(0,0) \cap (\Re \times [0,\infty[)\}$.

Here $\phi(x,t)$ is a test function with compact support on the boundary (i.e., $\phi(x,t)$ is zero outside the boundary). In the weak solution (2.76), the partial derivative is moved to the test function that is guaranteed to be smooth. In addition, the definition above is an extension of the classical solution according to

Theorem 2.6.2 *Suppose $\rho \in C^1(\Re \times [0,\infty[)$ is a classical solution of (2.72). Then ρ is also a weak solution.*

Proof is given in [86].

We have to keep in mind that a *weak solution* might not be a classical solution. So we need necessary and sufficient conditions for the weak solutions to be the correct solution. We start by the necessary condition for a piecewise-smooth weak solution known as the Rankine-Hugoniot condition given by

$$s = \frac{[f(\rho)]}{[\rho]} = \frac{f(\rho_R) - f(\rho_L)}{\rho_R, -\rho_L} \quad (2.77)$$

where s is the shock speed. Let's look at the earlier example in Fig. 2.8, where the initial density was given by

$$\rho(x,0) = \begin{cases} \frac{\rho_m}{2} & x < 0, \\ \rho_m & x \geq 0, \end{cases} \quad (2.78)$$

and we seek a solution to our Cauchy problem (2.72), using the method of characteristics. Since the characteristics do overlap at some point in time, the shock speed is calculated from (2.77)

$$\begin{aligned}
s &= \frac{(v_f \rho_R - v_f \rho_R^2/\rho_m) - (v_f \rho_L - v_f \rho_L^2/\rho_m)}{\rho_R - \rho_L} \\
&= v_f - \frac{v_f}{\rho_m}(\rho_R + \rho_L) \\
&= v_f - \frac{v_f}{\rho_m}\left(\rho_m + \frac{\rho_m}{2}\right) \\
&= -0.5\, v_f
\end{aligned}$$

2.6 Method of Characteristics

and the solution is given by

$$\rho(x,t) = \begin{cases} \frac{\rho_m}{2} & x < st, \\ \rho_m & x \geq st. \end{cases} \qquad (2.79)$$

This solution is shown in Fig. 2.10.

Let's look now at the example of Fig. 2.9, and try to solve the traffic flow problem there. We will give two methods to find the solution and discuss which one must be used to get the correct solution. Using the initial density values

$$\rho(x,0) = \begin{cases} \rho_m & x < 0, \\ \rho_m/2 & x \geq 0, \end{cases} \qquad (2.80)$$

we get the first solution as a shock wave given by

$$s = v_f - \frac{v_f}{\rho_m}(\rho_R + \rho_L) = v_f - \frac{v_f}{\rho_m}(\rho_m/2 + \rho_m) = -0.5\, v_f$$

As we can see "$-0.5\,v_f$" is the same shock speed as in the previous example and it is plotted in the Fig. 2.11b. The second solution is continuous and it provides another way to fill the wedge. It is called the *rarefaction wave* solution. The general form of the solution for traffic flow is given by

$$\rho(x,t) = \begin{cases} \rho_L & x < f'(\rho_L)\,t, \\ f'(\frac{x}{t})^{-1} & f'(\rho_L)\,t \leq x < f'(\rho_R)\,t, \\ \rho_R & x \geq f'(\rho_R)\,t, \end{cases} \qquad (2.81)$$

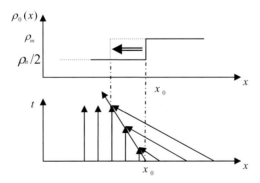

Fig. 2.10. Shock solution

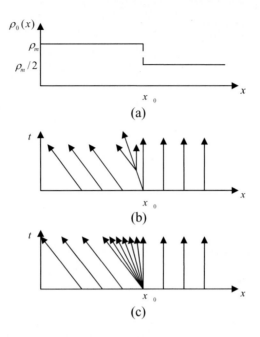

Fig. 2.11. Initial density profile followed by two weak solutions, shock and fan respectively

where the "−1" is an inverse mapping. For the traffic flow problem, $f'(\frac{x}{t})^{-1}$ can be found by letting

$$f'(\frac{x}{t})^{-1} = v_f - \frac{v_f \rho}{\rho_m} = \frac{x}{t}, \qquad (2.82)$$

and solving for the density solution to get

$$\rho(x,t) = f'(\frac{x}{t})^{-1} = \frac{\rho_m}{2} - \frac{\rho_m x}{2 v_f t}. \qquad (2.83)$$

Then, the continuous solution for the PDE is given by

$$\rho(x,t) = \begin{cases} \rho_L & x < -v_f t, \\ f'(\frac{x}{t})^{-1} & -v_f t \le x < 0, \\ \rho_R & x \ge 0. \end{cases} \qquad (2.84)$$

For the initial data given in Fig. 2.11a, the *rarefaction wave* solution is plotted in Fig. 2.11c.

2.6 Method of Characteristics

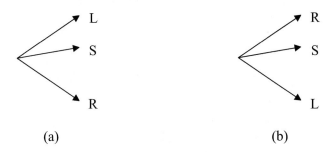

Fig. 2.12. Lax shock condition for traffic flow problem: (**a**) not a shock, (**b**) shock

Such multiplicity of solutions is unacceptable. Thus we need a selection criterion that picks out the physically reasonable solution from among the possible weak solutions (2.81), and (2.84). Lax shock condition or *Lax entropy* condition [66] is a sufficient condition for scalar conservation laws. The condition for the traffic flow PDE states that the discontinuous solution for the traffic flow problem is admissible (i.e., a shock solution is selected with the direction of increased entropy) if

$$\rho_L < s < \rho_R, \tag{2.85}$$

otherwise, the rarefaction wave solution is the admissible one (see Fig. 2.12 for easy interpretation). In the example of Fig. 2.11a, and according to the Lax condition, the rarefaction solution is the admissible one and the shock solution is not admissible. Finally, we summarize the solution for the LWR model by mentioning the following two points:

- The solution is piece-wise smooth as $t \mapsto \infty$ with jumps in density (shocks) separating the pieces.

- This means traffic is predicted to be stable with transition between stable regions approximated by discontinuous shocks.

Chapter 3

Crowd Models for 2-D

3.1 Introduction

In this chapter we preseodelnt four macroscopic crowd dynamic models that can be used to study crowd behavior. These models modify the 1-D traffic flow models so that bi-directional controlled flow is possible. The models have different characteristics; for instance, the first model is a simple conservation of pedestrians based on the one-equation partial differential equation model given in Sect. 2.5.1. The other three are based on a 1-D space systems of two-coupled partial deferential equations (two-equation models) given in Sects. 2.5.2, 2.5.3, and 2.5.4 respectively.

The first system model which is called "the second" model in this chapter follows fluid flow closely, such that it carries the isotropic nature of fluid flow. The last two models move away from fluid flow and mimic traffic flow observed behavior by having the anisotropic property. In addition, the last model not only inherits the same anisotropic property, but it can also be derived from microscopic *car-following* model [1].

This chapter is organized in the following matter. Section 3.2 gives the 2-D flow theory for hyperbolic PDE's. In Sect. 3.3 we present the one-equation crowd model. Section 3.4 shows the first system model derivation, conservation form, and eigenvalues followed by Sects. 3.5 and 3.6 for the second and third systems respectively.

A comparison between the crowd models developed based on the analytical results is presented in Sect. 3.7. Finally, we conclude this chapter with a linearized version of the four models.

3.2 Traffic Flow Theory in 2-D

In this section we extend the same ideas that is found in Sect. 2.6. We consider the scalar equation in 2-D:

$$\rho_t + f_x(\rho) + g_y(\rho) = 0 \quad in \ \Re^2 \times \Re^+, \tag{3.1}$$
$$\rho(.,0) = \rho_0 \quad in \ \Re^2, \tag{3.2}$$

for given functions f and g and ρ. We assume that f and g are sufficiently smooth. As we have seen the solution of the scalar 1-D conservation law can become singular within finite time, the same can happen here. Therefore, similar to the definition given by 2.6.2, a weak solution of (3.1), and (3.2) need to be defined.

Definition 3.2.1 Let $\rho_0 \in L^\infty(\Re^2)$. Then ρ is called a weak solution of (3.1), (3.2) if and only if $\rho \in L^\infty(\Re^2 \times \Re^+)$ and if,

$$\int_{\Re^2} \int_{\Re^+} [\rho \, \phi_t + f(\rho) \, \phi_x + g(\rho) \, \phi_y] + \int_{\Re^2} \rho_0 \, \phi(.,0) = 0$$

is satisfied for every $\phi \in C_0^\infty(\Re^2 \times [0,\infty[)$

For completion, we give the entropy condition for the 2-D one-equation case.

Definition 3.2.2 A weak solution of (3.1), (3.2) is called an entropy solution if we have for all $\phi \in C_0^\infty(\Re^2 \times \Re^+)$, $\phi \geq 0$ and for all $k \in \Re$

$$\int_{\Re^2} \int_{\Re^+} \{\partial_t \phi \, |\rho - k| \ + \ \partial_x \phi \, sign(\rho - k) \, [f(\rho) - f(k)]$$
$$+ \ \partial_y \phi \, sign(\rho - k) \, [g(\rho) - g(k)]\} \geq 0$$

The following theorem gives the existence and uniqueness of an entropy solution for the scalar 2-D hyperbolic PDE in (3.1),

Theorem 3.2.1 Let $\rho_0 \in L^1(\Re^2) \cap L^\infty(\Re^2)$. Then there exist one and only one entropy solution ρ of (3.1), (3.2) and $\rho \in C^0([0,T], L^1(\Re^2)) \cap L^\infty([0,T], \Re^2)$.

3.3 One Equation Crowd Model

The details of this proof can be found in [65]. For systems in 2-D, Let $Q \in \Re^m$, then the scalar equation (3.1) and (3.2) become a system of PDE's for 2-D given by

$$Q_t + F_x(Q) + G_y(Q) = 0 \quad in \ \Re^2 \times \Re^+, \tag{3.3}$$
$$Q(.,0) = Q_0 \quad in \ \Re^m, \tag{3.4}$$

and

Definition 3.2.3 The system is called (strictly) hyperbolic system if all eigenvalues of $\alpha F'(Q) + \xi G'(Q)$ are real (and distinct) for all $\alpha, \xi \in \Re, Q \in \Re^m$.

We also note that if the two matrices coefficients do not commute, i.e., $AB \neq BA$ and the matrices $A(Q)$ and $B(Q)$ do not have the same eigenvectors [67], then they can be diagonalized separately

$$A = R^x \Lambda (R^x)^{-1}, \quad B = R^y \Lambda (R^y)^{-1}, \tag{3.5}$$

where the two matrices have different eigenvectors. This means that the PDE's are more intricately coupled.

For 2-D hyperbolic PDE's, the general existence of a solution does not exist for (3.3) globally in time [13]. This implies that there are no convergent results for numerical schemes. Therefore we can only use some schemes that have been successful in numerical test problems as we will see in the next chapter [64].

3.3 One Equation Crowd Model

In this section, we present a 2-D scalar, nonlinear, time-varying, hyperbolic PDE crowd model. This is a simple extension to the LWR 1-D conservation of continuity model given in Sect. 2.4.3. The model consist of crowd concentration (density) and flow rate $f(\rho) = \rho V(\rho)$ in the x-axis, and $g(\rho) = \rho U(\rho)$ in the y-axis. We use the fundamental velocity–density relation given by (2.15) to describe $V(\rho)$ and $U(\rho)$. The free flow speed v_{f1}, v_{f2} are the velocities in the x-axis and y-axis, and they are used as control parameters to direct crowd flow to any desired direction. Here, they are taken as constant, but in control design, they become a function of time and space. The model in conservation form is given by

$$\rho_t + (\rho V(\rho))_x + (\rho U(\rho))_y = 0, \tag{3.6}$$

and its quasi-linear form is

$$\rho_t + f'(\rho)\rho_x + g'(\rho)\rho_y = 0. \tag{3.7}$$

The characteristic slopes, or the eigenvalues of this scalar PDE are simply $f'(\rho)$, and $g'(\rho)$ and they are given by

$$\lambda^f = v_{f_1}(1 - 2\rho/\rho_m), \tag{3.8}$$
$$\lambda^g = v_{f_2}(1 - 2\rho/\rho_m), \tag{3.9}$$

The eigenvalues are real and distinct, therefore, the LWR model is classified as strictly hyperbolic since $f'(\rho)$, and $g'(\rho)$ are distinct and real respectively. The model carries the anisotropic property as the maximum characteristics are v_{f1} and v_{f2} respectively.

3.4 First System Crowd Dynamic Model

The first system model to be extended from 1-D to 2-D is based on the higher order model given in Sect. 2.5.2. The model is classified as nonlinear, time-varying, hyperbolic PDE traffic flow system. This model uses two-coupled PDE's to describe crowd flow; the conservation of continuity and a second equation that looks like the momentum equation in fluid flow with a modification to the "pressure" term to mimic crowd motion. This model is known for its isotropic nature that is preserved when we extend the model to 2-D space assuming pedestrian motion is influenced from all directions.

3.4.1 Model Description

Here we extend (2.23), and (2.24) from 1-D to 2-D, and add relaxation terms to allow bi-directional flow. The first equation is the 2-D conservation law given by

$$\rho_t + (\rho v)_x + (\rho u)_y = 0 \tag{3.10}$$

where the density $\rho(x, y, t)$ depends on the two spatial dimensions and time, $v(x, y, t)$ and $u(x, y, t)$ are the x-axis and y-axis components of the velocity. The flux flow rate in both directions is represented by ρv and ρu. The above equation is defined on $x \in \Re$, $y \in \Re$ and time $t \in \Re^+$. We can derive the above equation from Fig. 3.1,

3.4 First System Crowd Dynamic Model

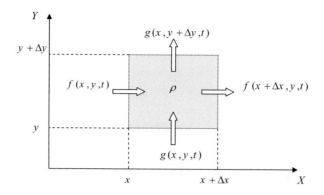

Fig. 3.1. Two-dimensional flow

where the conserved density changes according to the changes in the flow at the boundary endpoints.

The second equation is derived by taking a 2-D form of (2.24) and it is found to be

$$v_t + vv_x + uv_y + \frac{C(\rho)}{\rho}\rho_x = s_1, \qquad (3.11)$$

$$u_t + vu_x + uu_y + \frac{C(\rho)}{\rho}\rho_y = s_2, \qquad (3.12)$$

where the equations in non-conservative form represent the x-axis, and y-axis components. Initial condition are $\rho(x, y, 0) \geq 0$, and $\vec{v}(\vec{x}, 0) \leq |v_f|$. The "pressure" term is the same as in the 1-D case, and it is equal to $P(\rho) = \rho C(\rho)$, where $C(\rho)$ is the anticipation factor that is equal to a constant squared C_0^2. The relaxation terms

$$s_1 = \frac{V(\rho) - v}{\tau} \qquad (3.13)$$

$$s_2 = \frac{U(\rho) - u}{\tau} \qquad (3.14)$$

rule is to capture how pedestrians modify their velocity according to the desired one $V(\rho)$, and $U(\rho)$ as given in (2.15) over a period of time denoted by the relaxation time τ.

3.4.2 Conservation Form and Eigenvalues

The model developed is a nonlinear, time-varying hyperbolic PDE system. There is no analytical solution for this kind of systems, and we know for hyperbolic PDE's we have discontinuous solutions, i.e., more than one solution at some point in time and space (for more details on hyperbolic PDE's, see [67]). Therefore, we need to put the model in a form that can be solved numerically and provide results that are consistent with the observed behavior. To do so, we will derive the conservation form of the model which will be used in finding the numerical solution (more on this will be discussed in the next chapter). Using this form we also prove the system hyperbolic property and its isotropic nature by finding the system roots according to definition 3.2.3.

After some manipulations to the equations in (3.11), and (3.12) we rewrite them in conservation form as

$$(\rho v)_t + (\rho v^2 + \rho C_0^2)_x + (\rho v u)_y = \rho s_1, \tag{3.15}$$

$$(\rho u)_t + (\rho v u)_x + (\rho u^2 + \rho C_0^2)_y = \rho s_2. \tag{3.16}$$

In vector form, the model can be written as

$$Q_t + F(Q)_x + G(Q)_y = S \tag{3.17}$$

where Q is the conservative variables (states), F and G are the fluxes in the two space dimension, and S is considered as the source term. These are given by

$$Q = \begin{bmatrix} \rho \\ \rho v \\ \rho u \end{bmatrix}, \quad F(Q) = \begin{bmatrix} \rho v \\ \rho v^2 + C_0^2 \rho \\ \rho v u \end{bmatrix}, \quad G(Q) = \begin{bmatrix} \rho u \\ \rho v u \\ \rho u^2 + C_0^2 \rho \end{bmatrix}$$

$$S = \begin{bmatrix} 0 \\ \rho \left(\dfrac{V(\rho) - v}{\tau} \right) \\ \rho \left(\dfrac{U(\rho) - u}{\tau} \right) \end{bmatrix}.$$

Next we write the system in quasi-linear form, and by setting the relaxation terms to zero, our homogenous hyperbolic PDE in vector form is

$$Q_t + A(Q)\,Q_x + B(Q)\,Q_y = 0 \tag{3.18}$$

3.4 First System Crowd Dynamic Model

where the flux Jacobian matrices $A(Q)$ and $B(Q)$ can be found from the partial derivative of the fluxes given by

$$A = \frac{\partial F}{\partial Q} = \begin{bmatrix} 0 & 1 & 0 \\ C_0^2 - v^2 & 2v & 0 \\ -vu & u & v \end{bmatrix}, \quad B = \frac{\partial G}{\partial Q} = \begin{bmatrix} 0 & 0 & 1 \\ -vu & u & v \\ C_0^2 - u^2 & 0 & 2u \end{bmatrix}.$$

The eigenvalues of the matrices A and B are found from the flux Jacobian matrices roots calculated from

$$|A(Q) - \lambda I| = 0. \tag{3.19}$$

The corresponding eigenvalues and eigenvectors for the A matrix are given by

$$\lambda_1^A = v - C_0, \quad \lambda_2^A = v, \quad \lambda_3^A = v + C_0,$$

and

$$e_1^A = \begin{bmatrix} 1 \\ v - C_0 \\ u \end{bmatrix}, \quad e_2^A = \begin{bmatrix} 0 \\ 0 \\ 1 \end{bmatrix}, \quad e_3^A = \begin{bmatrix} 1 \\ v + C_0 \\ u \end{bmatrix}.$$

For the B matrix they are given by

$$\lambda_1^B = u - C_0, \quad \lambda_2^B = u, \quad \lambda_3^B = u + C_0,$$

and

$$e_1^B = \begin{bmatrix} 1 \\ v \\ u - C_0 \end{bmatrix}, \quad e_2^B = \begin{bmatrix} 0 \\ 1 \\ 0 \end{bmatrix}, \quad e_3^B = \begin{bmatrix} 1 \\ v \\ u + C_0 \end{bmatrix}.$$

To check if the system is (strictly) hyperbolic or not, we need to do more than just getting the eigenvalues for A and B separately. Since this is a 2-D problem, we find the overall system eigenvalues from the roots of the combined Jacobian matrixes satisfying definition 3.2.3. Hence, the first crowd dynamic model eigenvalues are found to be

$$\check{\lambda}_1 = \check{v} - C_0, \quad \check{\lambda}_2 = \check{v}, \quad \check{\lambda}_3 = \check{v} + C_0, \tag{3.20}$$

where $\check{v} = \alpha v(x,y,t) + \xi u(x,y,t)$. The corresponding eigenvectors are given by

$$e_1^{\check{A}} = \begin{bmatrix} 1 \\ v - \alpha C_0 \\ u - \xi C_0 \end{bmatrix}, \quad e_2^{\check{A}} = \begin{bmatrix} 0 \\ -\xi \\ \alpha \end{bmatrix}, \quad e_3^{\check{A}} = \begin{bmatrix} 1 \\ v + \alpha C_0 \\ u + \xi C_0 \end{bmatrix}.$$

The system eigenvalues are real and distinct, and their corresponding eigenvectors are linearly independent, which confirms that the model is strictly hyperbolic. In addition, the system preserves its isotropic nature as seen from $\lambda_{1,3}$, where one of them is always greater than λ_2. This means that information from all directions affect pedestrians' motion (isotropic nature) and this is the most distinct characteristic of this model.

3.5 Second Crowd Dynamic Model

Here we present a crowd dynamic model in 2-D space based on a 1-D traffic flow model given in [4, 84]. This model is considered an improvement over the first system model since it does not closely follow fluid and gas-like models. In addition, it allows observed traffic conditions such as traffic flow with the flow in front, rather than with the flow that is upstream. Hence, the model carries the desired anisotropic nature of traffic flow, which is carried from the 1-D model by modifying the "pressure" terms.

3.5.1 Model Description

This model is also a nonlinear, time-varying, hyperbolic system of two PDE's. We develop the model by first extending (2.23), and (2.39) from a 1-D to a 2-D PDE system. The first PDE is the same conservation of continuity given in (3.10). The second equation is derived for the x-axis and y-axis by applying the convective derivative (2.38) on the pressure terms for the 2-D case. By doing so, the resulting second equation is given by

$$(v + P_1(\rho, v))_t + v(v + P_1(\rho, v))_x + u(v + P_1(\rho, v))_y = \rho s_1, \quad (3.21)$$
$$(u + P_2(\rho, u))_t + v(u + P_2(\rho, u))_x + u(u + P_2(\rho, u))_y = \rho s_2. \quad (3.22)$$

3.5 Second Crowd Dynamic Model

Initial condition are $\rho(x,y,0) \geq 0$, and $v(x,0) \leq |v_{f1}|$ and $u(y,0) \leq |v_{f2}|$. The $P_1(\rho,v)$, and $P_2(\rho,u)$ functions proposed here are given by

$$P_1(\rho,v) = \frac{v\rho^{\gamma+1}}{\beta - \rho^{\gamma+1}}, \qquad (3.23)$$

$$P_2(\rho,u) = \frac{u\rho^{\gamma+1}}{\beta - \rho^{\gamma+1}}, \qquad (3.24)$$

and valid for $\gamma > 0$, and $\beta > \rho_m^{\gamma+1}$ such that we do not divide by zero or change the sign that could affect the system dynamics. We modified these functions from the 1-D case to maintain the increasing property of the density that is required to achieve the anisotropic property of the system. Thus, the model preserve the anisotropic property and the modification is justified by the system Eigenvalues found next. The second modification is adding the relaxation terms s_1, and s_2 that made possible to simulate bi-directional pedestrian motion.

3.5.2 Conservation Form and Eigenvalues

To show that the crowd dynamic model in (3.10), (3.21), and (3.22) is a hyperbolic PDE system, and numerically find its solution. We need to find the system conservation form and find its eigenvalues. We restrict our self to derive the x-component only since the y-component can be obtained by the same procedure. First, we start by multiplying (3.21) by ρ to get

$$\rho(v + P_1(\rho,v))_t + \rho v\,(v + P_1(\rho,v))_x + \rho u\,(v + P_1(\rho,v))_y = \rho s_1, \quad (3.25)$$

and we drop (ρ,v) from $P_1(\rho,v)$ for convenience for the rest of this section. We know from the product rule that

$$\begin{aligned}
(\rho\,(v+P_1))_t &= \rho_t(v+P_1) + \rho\,(v+P_1)_t, \\
(\rho v\,(v+P_1))_x &= (\rho v)_x(v+P_1) + (\rho v)(v+P_1)_x, \\
(\rho u\,(v+P_1))_y &= (\rho u)_y(v+P_1) + (\rho u)(v+P_1)_y,
\end{aligned}$$

and by substituting the above terms in (3.25) and using (3.10) to write the model in conservation form (follow same steps for the Y-component using P_2) we get

$$\begin{aligned}
\rho_t + (\rho v)_x + (\rho u)_y &= 0 & (3.26) \\
(\rho(v+P_1))_t + (\rho v\,(v+P_1))_x + (\rho u\,(v+P_1))_y &= \rho s_1 & (3.27) \\
(\rho\,(u+P_2))_t + (\rho v\,(u+P_2))_x + (\rho u\,(u+P_2))_y &= \rho s_2 & (3.28)
\end{aligned}$$

Next is to write the system in 2-D vector form (3.17) where Q is the conservative variables (states), and F & G are the fluxes in the x and y-directions respectively. The model in vector form is given by

$$\begin{bmatrix} \rho \\ \rho(v+P_1) \\ \rho(u+P_2) \end{bmatrix}_t + \begin{bmatrix} \rho v \\ \rho v(v+P_1) \\ \rho v(u+P_2) \end{bmatrix}_x + \begin{bmatrix} \rho u \\ \rho u(v+P_1) \\ \rho u(u+P_2) \end{bmatrix}_y = \begin{bmatrix} 0 \\ \rho s_1 \\ \rho s_2 \end{bmatrix}.$$
(3.29)

From the quasi-linear form (3.18), we can obtain A and B eigenvalues by finding the flux Jacobian matrices $A = \partial F/\partial Q$ and $B = \partial G/\partial Q$. For this model the matrices and the corresponding eigenvalues and eigenvector are found by first simplifying the calculations. To do so, we change variables by setting the states to be

$$Q = \begin{bmatrix} \rho \\ w \\ z \end{bmatrix}, \text{ and } \begin{cases} \rho(v+P_1) = w \Rightarrow v = \dfrac{w}{\rho} - P_1, \\ \rho(u+P_2) = z \Rightarrow u = \dfrac{w}{\rho} - P_2. \end{cases}$$

We substituted the new states in the fluxes to get

$$F(Q) = \begin{bmatrix} w - \rho P_1 \\ \dfrac{w^2}{\rho} - wP_1 \\ \dfrac{wz}{\rho} - zP_1 \end{bmatrix}, \quad G(Q) = \begin{bmatrix} z - \rho P_2 \\ \dfrac{wz}{\rho} - wP_2 \\ \dfrac{z^2}{\rho} - zP_2 \end{bmatrix},$$

and the Jacobian are

$$A(Q) = \begin{bmatrix} -P_1 - \rho P_{1\rho} & 1 - \rho P_{1w} & 0 \\ -\dfrac{w^2}{\rho^2} - wP_{1\rho} & 2\dfrac{w}{\rho} - P_1 - zP_{1w} & 0 \\ -\dfrac{wz}{\rho^2} - zP_{1\rho} & \dfrac{z}{\rho} - zP_{1w} & \dfrac{w}{\rho} - P_1 \end{bmatrix}$$
(3.30)

$$B(Q) = \begin{bmatrix} -P_2 - \rho P_{2\rho} & 0 & 1 - \rho P_{2z} \\ -\dfrac{wz}{\rho^2} - wP_{2\rho} & \dfrac{z}{\rho} - P_2 & \dfrac{w}{\rho} - wP_{2z} \\ -\dfrac{z^2}{\rho^2} - zP_{2\rho} & 0 & 2\dfrac{z}{\rho} - P_2 - zP_{2z} \end{bmatrix}.$$
(3.31)

3.5 Second Crowd Dynamic Model

We solve for the eigenvalues and eigenvectors for the $A(Q)$ matrix to get

$$\lambda_1^A = \frac{w - (\rho P_1 + \rho^2 P_{1\rho} + w\rho P_{1w})}{\rho} = v - v\frac{(\gamma+1)\rho^{\gamma+1}}{\beta - \rho^{\gamma+1}}$$
$$= v - (\gamma+1)P_1$$

$$\lambda_{2,3}^A = \frac{w - \rho P_1}{\rho} = v$$

$$e_1^A = \begin{bmatrix} \frac{\beta - \rho^{\gamma+1}}{\beta u} \\ \frac{v}{u} \\ 1 \end{bmatrix}, \quad e_2^A = \begin{bmatrix} \frac{(\beta - \rho^{\gamma+1})^2}{\beta v(\beta + \gamma\rho^{\gamma+1})} \\ 1 \\ 0 \end{bmatrix}, \quad e_3^A = \begin{bmatrix} 0 \\ 0 \\ 1 \end{bmatrix}.$$

and for $B(Q)$ matrix we obtain

$$\lambda_1^B = \frac{z - (\rho P_2 + \rho^2 P_{2\rho} + z\rho P_{2z})}{\rho} = u - u\frac{(\gamma+1)\rho^{\gamma+1}}{\beta - \rho^{\gamma+1}}$$
$$= u - (\gamma+1)P_2$$

$$\lambda_{2,3}^B = \frac{z - \rho P_2}{\rho} = u$$

$$e_1^B = \begin{bmatrix} \frac{\beta - \rho^{\gamma+1}}{\beta v} \\ 1 \\ \frac{u}{v} \end{bmatrix}, \quad e_2^B = \begin{bmatrix} 0 \\ 1 \\ 0 \end{bmatrix}, \quad e_3^B = \begin{bmatrix} \frac{(\beta - \rho^{\gamma+1})^2}{\beta u(\beta + \gamma\rho^{\gamma+1})} \\ 0 \\ 1 \end{bmatrix}.$$

Here we also find the eigenvalues by solving the combined Jacobian matrixes satisfying definition 3.2.3. For this system the eigenvalues are found to be

$$\check{\lambda}_1 = \check{v} - \check{v}\frac{(1+\gamma)\rho^{\gamma+1}}{\beta - \rho^{\gamma+1}} = \check{v} - (1+\gamma)\check{P}, \quad \check{\lambda}_{2,3} = \check{v}$$

and their corresponding eigenvectors are given by

$$e_1^{\check{A}} = \begin{bmatrix} \frac{\beta - \rho^{\gamma+1}}{\beta u} \\ \frac{v}{u} \\ 1 \end{bmatrix}, \quad e_2^{\check{A}} = \begin{bmatrix} \alpha\frac{(\beta - \rho^{\gamma+1})^2}{\beta \check{v}(\beta + \gamma\rho^{\gamma+1})} \\ 1 \\ 0 \end{bmatrix}, \quad e_3^{\check{A}} = \begin{bmatrix} \xi\frac{(\beta - \rho^{\gamma+1})^2}{\beta \check{v}(\beta + \gamma\rho^{\gamma+1})} \\ 0 \\ 1 \end{bmatrix}$$

Clearly, we have real eigenvalues with two of them repeated. Nevertheless, we are able to find linearly independent eigenvectors, and therefore the crowd dynamic model derived is a nonlinear hyperbolic system of PDE. In addition, the "pressure" term is an increasing function of density, and from the eigenvalue λ_1, we are sure to have the maximum wave speed to be $\check{v}$. Thus, the model has the desired anisotropic property (information that affect the current position depends on current $\check{\lambda}_{2,3}$, and ahead $\check{\lambda}_1$ information only).

3.6 Third Crowd Dynamic Model

The third crowd dynamic model presented here is a macroscopic model derived from a microscopic car-following model. The model is based a 1-D macroscopic traffic flow model given in Sect. 2.5.4. This crowd 2-D model is different from the other models because it has a direct microscopic-to-macroscopic link. The model is classified as a traffic flow nonlinear, time-varying, hyperbolic system of two partial differential equations. The anisotropic property is carried from the microscopic to the macroscopic model assuming that pedestrian motion is influenced mainly from current and front conditions.

Here we present the macroscopic model, and then we derive the macroscopic system from the microscopic model. Finally, we put the crowd model in conservation form and find its eigenvalues and eigenvectors.

3.6.1 Model Description

The first equation is the 2-D conservation of continuity that conserve mass (pedestrians) given by (3.10). The second equation is similar to the momentum equations in 2-D for comprisable flow with some manipulation to mimic crowd dynamics and it is given by

$$v_t + vv_x + uv_y + \rho V'(\rho)(v_x + u_y) = \frac{V(\rho) - v}{\tau} \quad (3.32)$$

$$u_t + vu_x + uu_y + \rho U'(\rho)(v_x + u_y) = \frac{U(\rho) - u}{\tau} \quad (3.33)$$

where $V(\rho)$ and $U(\rho)$ are the desired velocities functions meant to mimic pedestrian behavior given by the velocity–density relation

3.6 Third Crowd Dynamic Model

(2.15), and $\rho V'(\rho)$ is the traffic sound speed at which small traffic disturbances are propagated relative to the moving crowd stream. The relaxation terms $(V(\rho) - v)/\tau$ and $(U(\rho) - u)/\tau$ are added to the model as a modification to keep speed concentration in equilibrium, where τ is this process relaxation time. They are important to the system because v_{f1} and v_{f2} can be used as control parameters for crowd speed and direction. Initial conditions are $\rho(x, y, 0) \geq 0$, $v(x, 0) \leq |v_{f1}|$, and $u(y, 0) \leq |v_{f2}|$.

3.6.2 Derivation of a Macroscopic Model from a Microscopic Model in 2-D

In this section we will show how to derive the second equation of the model (3.32), (3.33) from the microscopic *car-following* model that is used to represent traffic flow in 1-D. For 2-D we use the same idea given in Chap. 2 by Sect. 2.5.4 and start the derivation from a microscopic model in 2-D given by

$$\tau(\mathbf{s}_n(t)) \ddot{\mathbf{x}}_n(t) = \dot{\mathbf{x}}_{n+1}(t) - \dot{\mathbf{x}}_n(t) + A \left[V \left(\frac{\Delta X}{\mathbf{s}_n(t)} \right) - \dot{\mathbf{x}}_n(t) \right], \quad (3.34)$$

where

$$\mathbf{s}_n(t) = \mathbf{x}_{n+1}(t) - \mathbf{x}_n(t), \quad (3.35)$$

is a function of the local spacing between pedestrians, $\mathbf{x}_n(t)$ is the 2-D position of the nth pedestrian, ΔX is the width of a single pedestrian, $\ddot{\mathbf{x}}_n(t)$ is the acceleration, $\dot{\mathbf{x}}_n(t)$ is the velocity, and $\tau(\mathbf{s}_n(t))$ is the pedestrian response (relaxation) time to the headway distance and it depends on spacing. This response time is different than pedestrian reaction time which is on average a small constant number. For the constant $A > 0$ a relaxation term is added and for the *homogeneous* case we set $A = 0$. Using the above notations, we rewrite (3.34) for the x-component where the local spacing s is in the x-direction only. We start by defining the velocity field $v(x, y, t) : \dot{x}(t) = v(x_n(t), t)$, and pedestrian spacing function $s(x, t) : s_n(t) = s(x_n(t), t)$. We also define local density by

$$\rho(x, y, t) : \rho_n(t) = \frac{\Delta X}{s_n(t)}, \quad (3.36)$$

which is the number of people per unit length. In our definition the density is normalized and therefore dimensionless, so that jam density (maximum capacity) is $\rho_m = 1$. We substitute the new variables in (3.34) for the x-component and get

$$\tau(s(x(t),t))\frac{dv(x,y,t)}{dt} = \frac{d(s(x(t),t))}{dt} + A\left[V(\rho(x,t)) - v(x,y,t)\right]. \tag{3.37}$$

Using the convective derivative $\partial_t + v\partial_x + u\partial_y$ on the velocity component of the x-axis (for y-axis use u instead of v), we obtain

$$\tau(s)(v_t + vv_x + uv_y) = (s_t + vs_x + us_y) + A\left[V(\rho(x,y,t)) - v(x,y,t)\right]. \tag{3.38}$$

From the conservation law (3.10), let $\rho = 1/s$, and by using the full derivative

$$D_x(\frac{a}{b}) = \frac{bD_x a - aD_x b}{b^2}, \tag{3.39}$$

we get

$$s_t + vs_x + us_y = sv_x + su_y, \tag{3.40}$$

and by direct substituting in the right hand side of (3.38), we arrive to our desired equation

$$(v_t + vv_x + uv_y) = \frac{s}{\tau(s)}(v_x + u_y) + \frac{A}{\tau(s)}\left[V(\rho(x,y,t)) - v(x,y,t)\right], \tag{3.41}$$

where

$$\frac{s}{\tau(s)} = -C(\rho) = -\rho V'(\rho) \geq 0, \tag{3.42}$$

is the sound wave speed, and A is a constant equals one. Finally, the process relaxation term τ replaces $\tau(s)$ for macroscopic behavior to get (3.32). This completes the derivation of the 2-D macroscopic model from its microscopic counterpart for the x-axis. We follow similar steps for the y-axis. The macroscopic conservative form of this model for the x-axis is derived next.

3.6.3 Conservation Form and Eigenvalues

To find the eigenvalues of the system and check if the system is hyperbolic (real eigenvalues), we write the model in conservation vector form. The same form will be used in the numerical simulation

3.6 Third Crowd Dynamic Model

to obtain the system response. We start deriving the conservation form by expanding the derivatives in (3.10) to get

$$\rho(v_x + u_y) = -(\rho_t + \rho_x v + \rho_y u), \tag{3.43}$$

and we know that for $V(\rho)$ given by (2.15)

$$V'(\rho)\rho_t = V_t(\rho), \quad V'(\rho)\rho_x = V_x(\rho), \quad V'(\rho)\rho_y = V_y(\rho). \tag{3.44}$$

Then substitute the above in (3.32) and multiply by ρ to get

$$\rho(v - V(\rho))_t + \rho v \left(v - V(\rho)\right)_x + \rho u \left(v - V(\rho)\right)_y = \rho \frac{V(\rho) - v}{\tau}, \tag{3.45}$$

and by using the product rules

$$\begin{aligned}
(\rho(v - V(\rho)))_t &= \rho_t (v - V(\rho)) + \rho(v - V(\rho))_t, & (3.46) \\
(\rho v (v - V(\rho)))_x &= (\rho v)_x (v - V(\rho)) + (\rho v)(v - V(\rho))_x, & (3.47) \\
(\rho u (v - V(\rho)))_y &= (\rho u)_y (v - V(\rho)) + (\rho u)(v - V(\rho))_y, & (3.48)
\end{aligned}$$

we substitute in (3.45) and use the conservation law (3.10) to write the model in conservation form (we follow similar steps for the y-component using $U(\rho)$ instead of $V(\rho)$). Finally, after manipulating (3.41) we get our equation in conservative form as

$$(\rho(v - V(\rho)))_t + (\rho v \left(v - V(\rho)\right))_x + (\rho u \left(v - V(\rho)\right))_y = \rho \frac{V(\rho) - v}{\tau}, \tag{3.49}$$

and for the y-direction it is given by

$$(\rho (u - U(\rho)))_t + (\rho v (u - U(\rho)))_x + (\rho u (u - U(\rho)))_y = \rho \frac{U(\rho) - u}{\tau}. \tag{3.50}$$

Next we write the system in 2-D vector form (3.17), where Q is the conservative variables (states), F and G are the fluxes in the x and y-directions respectively, and S can be considered as the source term. These are given by

$$\begin{bmatrix} \rho \\ \rho(v - V(\rho)) \\ \rho(u - U(\rho)) \end{bmatrix}_t + \begin{bmatrix} \rho v \\ \rho v (v - V(\rho)) \\ \rho v (u - U(\rho)) \end{bmatrix}_x + \begin{bmatrix} \rho u \\ \rho u (v - V(\rho)) \\ \rho u (u - U(\rho)) \end{bmatrix}_y = S \tag{3.51}$$

and $S = [0 : s_1 : s_2]^T$. Next we rewrite the system in the general quasi-linear form given by (3.18), where the source term is zero and the flux Jacobian matrices are found from $A(Q) = \partial F(Q)/\partial Q$, and $B(Q) = \partial G(Q)/\partial Q$. For this system the matrices and their corresponding eigenvalues and eigenvectors are found by first setting the conservative values (states) as

$$Q = \begin{bmatrix} \rho \\ w \\ z \end{bmatrix}, \text{ where } \begin{cases} \rho(v - V(\rho)) = w \Rightarrow v = \dfrac{w}{\rho} + V(\rho) \\ \rho(u - U(\rho)) = z \Rightarrow u = \dfrac{w}{\rho} + U(\rho) \end{cases}$$

We rewrite the fluxes $F(Q)$ and $G(Q)$ as

$$F(Q) = \begin{bmatrix} w + \rho V(\rho) \\ \dfrac{w^2}{\rho} + wV(\rho) \\ \dfrac{wz}{\rho} + zV(\rho) \end{bmatrix}, \quad G(Q) = \begin{bmatrix} z + \rho U(\rho) \\ \dfrac{wz}{\rho} + wU(\rho) \\ \dfrac{z^2}{\rho} + zU(\rho) \end{bmatrix}.$$

Their corresponding Jacobian matrices are found to be

$$A(Q) = \begin{bmatrix} V(\rho) + \rho V'(\rho) & 1 & 0 \\ -\dfrac{w^2}{\rho^2} + wV'(\rho) & 2\dfrac{w}{\rho} + V(\rho) & 0 \\ -\dfrac{wz}{\rho^2} + zV'(\rho) & \dfrac{z}{\rho} & \dfrac{w}{\rho} + V(\rho) \end{bmatrix}, \quad (3.52)$$

$$B(Q) = \begin{bmatrix} U(\rho) + \rho U'(\rho) & 0 & 1 \\ -\dfrac{wz}{\rho^2} + wU'(\rho) & \dfrac{z}{\rho} + U(\rho) & \dfrac{w}{\rho} \\ -\dfrac{z^2}{\rho^2} + zU'(\rho) & 0 & 2\dfrac{z}{\rho} + U(\rho) \end{bmatrix}. \quad (3.53)$$

Solving for the eigenvalues and eigenvectors of $A(Q)$ we get

$$\lambda_1^A = v + \rho V'(\rho), \quad \& \quad \lambda_{2,3}^A = v, \quad (3.54)$$

$$e_1^A = \begin{bmatrix} 1 \\ v - V \\ u - U \end{bmatrix}, \quad e_2^A = \begin{bmatrix} 1 \\ v - V - \rho V' \\ 0 \end{bmatrix}, \quad e_3^A = \begin{bmatrix} 0 \\ 0 \\ 1 \end{bmatrix}$$

3.6 Third Crowd Dynamic Model

and for the $B(Q)$ matrix we obtain

$$\lambda_1^B = u + \rho U'(\rho), \quad \& \quad \lambda_{2,3}^B = u \tag{3.55}$$

$$e_1^B = \begin{bmatrix} 1 \\ v - V(\rho) \\ u - U(\rho) \end{bmatrix}, \quad e_2^B = \begin{bmatrix} 0 \\ 1 \\ 0 \end{bmatrix}, \quad e_3^B = \begin{bmatrix} 1 \\ 0 \\ u - U(\rho) - \rho U'(\rho) \end{bmatrix}$$

Since the system is a 2-D problem, and in order to verify that the system is hyperbolic, we need to check that our eigenvalues found earlier are valid for any combination of the roots of the combined system as defined by Sect. 3.2.3. The eigenvalues are found to be

$$\lambda_1 = \check{v} + \rho \check{V}', \quad \lambda_{2,3} = \check{v} \tag{3.56}$$

where $\check{v} = \alpha v(x,y,t) + \xi u(x,y,t)$, and $\check{V}' = \alpha V' + \xi U'$. Since the eigenvalues above are real, we conclude that our model is hyperbolic, and the fact that we also have repeated eigenvalues our model is not strictly hyperbolic. The first eigenvalue is always less or equal than the second one due to $\rho V'$ effect ($V' \leq 0$). From this fact we acknowledge the anisotropic property of the system that shows information can not travel faster than the actual wave, i.e., pedestrian movement is influenced from current and front stimuli only. Secondly, despite repeated eigenvalues, each matrix has linearly independent eigenvectors.

For each eigenvalue the corresponding eigenvectors are given by

$$e_1^{\check{A}} = \begin{bmatrix} 1 \\ v - V(\rho) \\ u - U(\rho) \end{bmatrix}, \quad e_2^{\check{A}} = \begin{bmatrix} 1 \\ \dfrac{\check{v} - \check{V}(\rho) - \rho \check{V}(\rho)}{\alpha} \\ 0 \end{bmatrix} \quad or \quad \begin{bmatrix} 0 \\ 1 \\ -\dfrac{\alpha}{\xi} \end{bmatrix}$$

$$e_3^{\check{A}} = \begin{bmatrix} 0 \\ -\dfrac{\xi}{\alpha} \\ 1 \end{bmatrix} \quad or \quad \begin{bmatrix} 1 \\ 0 \\ \dfrac{\check{v} - \check{V}(\rho) - \rho \check{V}(\rho)}{\xi} \end{bmatrix}.$$

For $e_2^{\check{A}}$ the "or" is to get back both e_2^A and e_2^B, and the same can be said about $e_3^{\check{A}}$.

3.7 Comparison Between the Models

Four models that are aimed at studying crowd dynamics have been presented in this chapter. They are derived from the 1-D traffic flow theory, with the proper adjustment to account for the bi-directional pedestrian flow. Table 3.1 summarizes the main characteristics of the four models.

We divide the models into two types; crowd model that closely follow gas and fluid-like traffic flow, and models that do not. We further divided the models into two types; scalar and systems of PDE's as shown in Fig. 3.2. The models are classified as nonlinear, hyperbolic, time-varying partial differential equation(s) with distinct characteristics separating them; for example, the first crowd model use Greenshield's equation (2.15) to describe the velocity as a function of density. This one-equation model is an extension to the scalar LWR traffic 1-D model. Although it is simple compared to the first model, it gives the desired anisotropic nature of traffic flow. In addition, the model is strictly hyperbolic. The second model is a system model that use two-coupled PDE's with an anticipation term $C(\rho)$ and the relaxation terms $\check{s} = \check{V}(\rho) - \check{v}/\tau$. These terms change the momentum equation to mimic crowd dynamic flow. The anticipation factor's role is to find the macroscopic response of pedestrian to traffic density, i.e., its response when interaction between pedestrians and obstacles are overlooked within their domain. On the other hand, the relaxation factor's role is to keep the velocity in equilibrium. If a pedestrian velocity is greater than the preferred value, then this relaxation term will slow the pedestrian traffic down. From the system eigenvalues (real and distinct), we can conclude that this model has an isotropic nature since one of its eigenvalues is always moving faster than the velocity itself. This means that information can travel faster than the flow, which may contradicts the observed behavior for pedestrians in normal situations, where pedestrians movement are mostly influenced from current and front conditions and not from behind. On the other hand, in panic situations like in the case of an emergency situations such as evacuation, and at exits we know that when crowds or pedestrians are moving close to each other there will be various influences from those surrounding them to some extent, including the ones from behind [46]. In the third model, we do have

3.7 Comparison Between the Models

Table 3.1. Comparison between the four models

Models Characteristics	One-Equation Model	First System	Second System	Third Model
Conservative Variables Q	ρ	$\rho, \rho v, \rho u$	$\rho, \rho(v+P_1), \rho(u+P_2)$	$\rho, \rho(v-\rho\mathcal{V}'), \rho(u-\rho U')$
"Pressure" term P	No	$\rho C(\rho) = \rho C_0^2$	$\dfrac{\vec{v}\rho^{\gamma+1}}{\beta - \rho^{\gamma+1}}, \gamma > 0, \beta > \rho_m^{\gamma+1}$	$-\rho\mathcal{V}'(\rho) = \rho\dfrac{v_f}{\rho_m}$
X - Direction $F(Q)$ λ_1	$f'(\rho) = v_{f1}(1 - 2\rho/\rho_m)$	$v - C_0$	$v - (\gamma+1)P_1(\rho)$	$v + \rho\mathcal{V}'$
Eigenvalues λ_2	No	v	v	v
λ_3	No	$v + C_0$	v	v
Y - Direction $G(Q)$ λ_1	$g'(\rho) = v_{f2}(1 - 2\rho/\rho_m)$	$u - C_0$	$u - (\gamma+1)P_2(\rho)$	$u + \rho\mathcal{V}'$
Eigenvalues λ_2	No	u	u	u
λ_3	No	$u + C_0$	u	u
PDE Type	Strictly Hyperbolic	Strictly Hyperbolic	Hyperbolic	Hyperbolic
Isotropic	No	Yes	No	No
Anisotropic	Yes	No	Yes	Yes

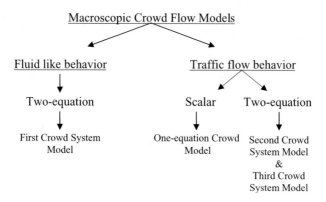

Fig. 3.2. Crowd model types

a relaxation term, but instead of the anticipation term that depends on space only in the first system model, a convective derivative on the "pressure" term is used. This change will enable the second PDE equation to predict the expected response of crowd behavior as time and space changes. The third system model has three real and two repeated eigenvalues; nevertheless, we found linearly independent eigenvectors that enabled us to solve the problem using numerical finite volume methods. The system has an anisotropic nature that is evident from its eigenvalues, where $\lambda_1 \leq \lambda_{2,3}$ for both velocity directions. This means that all information moves at a speed equal or less than the velocity of the corresponding state. This feature is preferred for traffic flow in general where pedestrian movement response is mostly due to conditions ahead like in normal conditions. In addition, we can relate a panic crowd situation to this from the fact that in most panic situations with high-density crowds, pedestrians tend to slowly form groups and move together without looking back. They just follow, assuming that the leader of this group knows or sees the way out.

The fourth model can be derived from a microscopic model, i.e., has a direct micro-to-macro link. In the model second PDE, the use of the convective derivative for the velocity terms leads to the sound wave speed $-C(\rho) = -\rho V'(\rho)$, which enables the model to predict the expected response of crowd behavior. That is they are used as anticipation terms to find the macroscopic response of pedestrians to traffic density. The terms $V(\rho(x, y, t))$, and $U(\rho(x, y, t))$ change the momentum equation to mimic crowd dynamic flow. At the same time, these terms help to keep velocity in equilibrium. If a pedestrian velocity is greater than the preferred value, these terms will tend to reduce the velocity and vice versa. The source terms in this model are essential in providing controlled behavior. They are considered as relaxation terms, where the change in direction would not be possible if they are not present.

The model has six real eigenvalues; three for each direction and two of them are repeated pair. Although we have repeated eigenvalues, the corresponding eigenvectors are linearly independent. This enabled us to perform simulations using numerical finite volume methods. The system has an anisotropic nature that appears from its eigenvalues $\lambda_1 \leq \lambda_{2,3}$.

3.8 Linearization

In this section we will linearize the PDE's modeling the crowd flow dynamics. We use Taylor series expansion to find the linearized models around some solution that we assume it exist and a perturbation, and we ignore higher order terms. Taylor series is given by

$$f(x + \Delta x) = f(x) + \Delta x f'(x) + \frac{(\Delta x)^2}{2!} f''(x) + \ldots \quad (3.57)$$

and for two variables Taylor series it becomes

$$f(x + \Delta x, y + \Delta y) = f(x, y) + [f_x(x, y) \Delta x + f_y(x, y) \Delta y] + H.O.T. \quad (3.58)$$

3.8.1 One Equation Crowd Model

For the one equation model given by

$$\rho_t + (\rho V(\rho))_x + (\rho U(\rho))_y = 0 \quad (3.59)$$

where $V(\rho)$ and $U(\rho)$ are the velocity–density functions given in (2.15). To linearize this model, we first assume that we have a density solution ρ_0 and vary it with small perturbation $\bar{\rho}$ to get

$$\rho = \rho_0 + \bar{\rho}.$$

We first need to find the velocity–density expansion to make our later substitutions easier. The expansion gives

$$V(\rho_0 + \bar{\rho}) = v_{f1}\left(1 - \frac{\rho_0 + \bar{\rho}}{\rho_m}\right) = v_{f1}\left(1 - \frac{\rho_0}{\rho_m}\right) - \frac{v_{f1}}{\rho_m}\bar{\rho}$$

$$= V(\rho_0) + V'\bar{\rho}, \qquad (3.60)$$

$$U(\rho_0 + \bar{\rho}) = U(\rho_0) + U'\bar{\rho}, \qquad (3.61)$$

where

$$V'(\rho) = \frac{\partial V(\rho)}{\partial \rho} = -\frac{v_{f1}}{\rho_m} = V', \qquad (3.62)$$

$$U'(\rho) = \frac{\partial U(\rho)}{\partial \rho} = -\frac{v_{f2}}{\rho_m} = U'. \qquad (3.63)$$

We consider $V(\rho_0)$, and $U(\rho_0)$ to be the velocity solutions, and in the subsequence sections when we have velocity dynamics it is equal to v_0, and u_0.

By substitution the above in our model and using Taylor series we get

$$(\rho_0 + \bar{\rho})_t + ((\rho_0 + \bar{\rho})(V(\rho_0) + V'\bar{\rho}))_x$$
$$+ ((\rho_0 + \bar{\rho})(U(\rho_0) + U'\bar{\rho}))_y = 0. \qquad (3.64)$$

Since ρ_0 is constant, all the derivatives associated with it are equal to zero. In addition, any high order terms are ignored (e.g. $\bar{\rho}\bar{\rho} = 0$) and the following linearized model is obtained

$$\bar{\rho}_t + V(\rho_0)\bar{\rho}_x + \rho_0 V'\bar{\rho}_x + U(\rho_0)\bar{\rho}_y + \rho_0 U'\bar{\rho}_y = 0. \qquad (3.65)$$

If we collect terms and rearrange the above equation as shown

$$\bar{\rho}_t + \underbrace{(V(\rho_0) + \rho_0 V')}_{f'(q)_{q=q_0} = v_{f1}\left(1 - \frac{2\rho_0}{\rho_m}\right)} \bar{\rho}_x + \underbrace{(U(\rho_0) + \rho_0 U')}_{g'(q)_{q=q_0} = v_{f2}\left(1 - \frac{2\rho_0}{\rho_m}\right)} \bar{\rho}_y = 0, \qquad (3.66)$$

where $q = \rho V(\rho)$ is the crowd flow, $f'(q) = \frac{\partial f(q)}{\partial \rho}$, and $g'(q) = \frac{\partial g(q)}{\partial \rho}$ are the characteristics speeds (also known as the eigenvalues). Thus, we end up with the same quasi-linear form that was introduced in (3.7), but here the eigenvalues are constant and the final form of the linearized model is given by

$$\bar{\rho}_t + f'(q)_{q=q_0}\bar{\rho}_x + g'(q)_{q=q_0}\bar{\rho}_y = 0. \qquad (3.67)$$

3.8 Linearization

3.8.2 First System Model

In this section we linearize the first system crowd dynamic model given by (3.10), (3.11), and (3.12). As we have done in the previous section, we use Taylor series expansion around an assumed solution trajectory. The difference here is because we linearize around the solutions of the density, and velocity in both directions as given by

$$\rho = \rho_0 + \bar{\rho}, \quad v = v_0 + \bar{v}, \quad u = u_0 + \bar{u},$$

where $v_0 = V(\rho_0)$, and $u_0 = U(\rho_0)$ are the assumed velocity solutions. We substitute the above into our model to get

$$(\rho_0 + \bar{\rho})_t + ((\rho_0 + \bar{\rho})(v_0 + \bar{v}))_x + ((\rho_0 + \bar{\rho})(u_0 + \bar{u}))_y = 0 \quad (3.68)$$

$$(v_0 + \bar{v})_t + (v_0 + \bar{v})(v_0 + \bar{v})_x + (u_0 + \bar{u})(v_0 + \bar{v})_y$$
$$+ \frac{C_0^2}{(\rho_0 + \bar{\rho})}(\rho_0 + \bar{\rho})_x = s_1 \quad (3.69)$$

$$(u_0 + \bar{u})_t + (u_0 + \bar{u})(u_0 + \bar{u})_x + (v_0 + \bar{v})(u_0 + \bar{u})_y$$
$$+ \frac{C_0^2}{(\rho_0 + \bar{\rho})}(\rho_0 + \bar{\rho})_y = s_2 \quad (3.70)$$

where

$$s_1 = \frac{(V(\rho_0 + \bar{\rho}) - (v_0 + \bar{v}))}{\tau}, \quad s_2 = \frac{(U(\rho_0 + \bar{\rho}) - (u_0 + \bar{u}))}{\tau}.$$

After terms cancelation we get

$$\bar{\rho}_t + v_0 \bar{\rho}_x + u_0 \bar{\rho}_y + \rho_0 (\bar{v}_x + \bar{u}_y) = 0, \quad (3.71)$$

$$\bar{v}_t + v_0 \bar{v}_x + u_0 \bar{v}_y + \frac{C_0^2}{\rho_0} \bar{\rho}_x = \frac{V'\bar{\rho} - \bar{v}}{\tau}, \quad (3.72)$$

$$\bar{u}_t + v_0 \bar{u}_x + u_0 \bar{u}_y + \frac{C_0^2}{\rho_0} \bar{\rho}_y = \frac{U'\bar{\rho} - \bar{u}}{\tau}, \quad (3.73)$$

which is the final form of the linearized model, and in vector form it is

$$Q_t + AQ_x + BQ_y = S(Q). \quad (3.74)$$

The matrices A and B are constant, and the vector $S(Q)$ contains the relaxation terms which are functions of the states. The vector form of the model is given by

$$\begin{bmatrix} \bar{\rho} \\ \bar{v} \\ \bar{u} \end{bmatrix}_t + \begin{bmatrix} v_0 & \rho_0 & 0 \\ C_0^2/\rho_0 & v_0 & 0 \\ 0 & 0 & v_0 \end{bmatrix} \begin{bmatrix} \bar{\rho} \\ \bar{v} \\ \bar{u} \end{bmatrix}_x$$
$$+ \begin{bmatrix} u_0 & 0 & \rho_0 \\ 0 & u_0 & 0 \\ C_0^2/\rho_0 & 0 & u_0 \end{bmatrix} \begin{bmatrix} \bar{\rho} \\ \bar{v} \\ \bar{u} \end{bmatrix}_y = S(Q), \quad (3.75)$$

where

$$S(Q) = \begin{bmatrix} 0 \\ (V'\bar{\rho} - \bar{v})/\tau \\ (U'\bar{\rho} - \bar{u})/\tau \end{bmatrix}$$

The eigenvalues of the system above are the same ones found in Sect. 3.4.2 for the nonlinear model, and they are given by

$$\lambda_1^A = v_0, \quad \lambda_{2,3}^A = v_0 \pm C_0, \quad \text{and} \quad \lambda_1^B = u_0, \quad \lambda_{2,3}^B = u_0 \pm C_0$$

We need to rewrite $S(Q)$ to separate the control variables for control design by the following manner

$$S(Q) = \underbrace{N(Q)}_{EQ} + M(Q)\eta$$

where η denote the controlled variables and E is a constant matrix. The above equation is given by

$$S(Q) = \begin{bmatrix} 0 & 0 & 0 \\ 0 & -1/\tau & 0 \\ 0 & 0 & -1/\tau \end{bmatrix} \begin{bmatrix} \bar{\rho} \\ \bar{v} \\ \bar{u} \end{bmatrix}$$
$$+ \begin{bmatrix} 0 & 0 & 0 \\ 0 & -\bar{\rho}/(\rho_m \tau) & 0 \\ 0 & 0 & -\bar{\rho}/(\rho_m \tau) \end{bmatrix} \begin{bmatrix} 0 \\ v_{f1} \\ v_{f2} \end{bmatrix}.$$

3.8.3 Second System Model

Here we linearize the second system model given by (3.10), (3.21), and (3.22). To start, we first linearize the "pressure" terms using

3.8 Linearization

Taylor series expansion given by (3.58) on the x-axis velocity component use u for y-axis given by

$$P(\rho, v) = \left(\frac{v \rho^{\gamma+1}}{\beta - \rho^{\gamma+1}} \right),$$

and we get

$$
\begin{aligned}
P(\rho_0 + \bar{\rho}, v_0 + \bar{v}) &= P(\rho_0, v_0) + [P_{\rho_0}(\rho_0, v_0)\bar{\rho} + P_{v_0}(\rho_0, v_0)\bar{v}] \\
&= \left(\frac{v_0 \rho_0^{\gamma+1}}{\beta - \rho_0^{\gamma+1}} \right) + \left[\frac{\beta(\gamma+1)v_0\rho_0^{\gamma+1}}{\left(\beta - \rho_0^{\gamma+1}\right)^2} \bar{\rho} + \frac{\rho_0^{\gamma+1}}{\beta - \rho_0^{\gamma+1}} \bar{v} \right] \\
&= P(\rho_0, v_0) + \underbrace{P_\rho(\rho, v)_{\rho_0, v_0}}_{\text{call it } P_\rho(\rho_0, v_0)} \bar{\rho} + \underbrace{P_v(\rho, v)_{\rho_0, v_0}}_{\text{call it } P_v(\rho_0, v_0)} \bar{v}.
\end{aligned}
$$

The linearized conservation of mass is given by (3.71). The linearized form for (3.21) and (3.22) is given by

$$
\begin{aligned}
\bar{v}_t &+ (P_\rho(\rho_0, v_0)\bar{\rho} + P_v(\rho_0, v_0)\bar{v})_t + v_0 \bar{v}_x \\
&+ v_0(P_\rho(\rho_0, v_0)\bar{\rho} + P_v(\rho_0, v_0)\bar{v})_x \\
&+ u_0 \bar{v}_y + u_0(P_\rho(\rho_0, v_0)\bar{\rho} + P_v(\rho_0, v_0)\bar{v})_y = \frac{V'\bar{\rho} - \bar{v}}{\tau} \quad (3.76)
\end{aligned}
$$

$$
\begin{aligned}
\bar{u}_t &+ (P_\rho(\rho_0, u_0)\bar{\rho} + P_u(\rho_0, u_0)\bar{u})_t + v_0 \bar{u}_x \\
&+ v_0(P_\rho(\rho_0, u_0)\bar{\rho} + P_u(\rho_0, u_0)\bar{u})_x \\
&+ u_0 \bar{u}_y + u_0(P_\rho(\rho_0, u_0)\bar{\rho} + P_u(\rho_0, u_0)\bar{u})_y = \frac{U'\bar{\rho} - \bar{u}}{\tau} \quad (3.77)
\end{aligned}
$$

In vector form $DQ_t + AQ_x + BQ_y = S$, we obtain

$$Q = \begin{bmatrix} \bar{\rho} \\ \bar{v} \\ \bar{u} \end{bmatrix}$$

$$D = \begin{bmatrix} 1 & 0 & 0 \\ P_\rho(\rho_0, v_0) & 1 + P_v(\rho_0, v_0) & 0 \\ P_\rho(\rho_0, u_0) & 0 & 1 + P_u(\rho_0, v_0) \end{bmatrix}$$

$$A = \begin{bmatrix} v_0 & \rho_0 & 0 \\ v_0 P_\rho(\rho_0, v_0) & v_0(1 + P_v(\rho_0, v_0)) & 0 \\ v_0 P_\rho(\rho_0, u_0) & 0 & v_0(1 + P_u(\rho_0, v_0)) \end{bmatrix}$$

$$B = \begin{bmatrix} u_0 & 0 & \rho_0 \\ u_0 P_\rho(\rho_0, v_0) & u_0\left(1 + P_u(\rho_0, u_0)\right) & 0 \\ u_0 P_\rho(\rho_0, v_0) & 0 & u_0\left(1 + P_u(\rho_0, u_0)\right) \end{bmatrix}$$

$$S = \begin{bmatrix} 0 \\ (V'\bar{\rho} - \bar{v})/\tau \\ (U'\bar{\rho} - \bar{u})/\tau \end{bmatrix}.$$

To put the system in the general vector form (3.74), we multiply the equation by the inverse of the nonsingular matrix D to get

$$Q_t + \left(D^{-1}A\right) Q_x + \left(D^{-1}B\right) Q_y = \left(D^{-1}S\right) \qquad (3.78)$$

The eigenvalues found from (3.78) are the same as the ones found in Sect. 3.5.2, but here they are time-invariant given by

$$\lambda_{1,2}^A = v_0, \quad \lambda_{2,3}^A = v_0 \left(1 - \frac{(\gamma + 1)\rho^{\gamma+1}}{\beta - \rho^{\gamma+1}}\right)$$

and

$$\lambda_{1,2}^B = u_0, \quad \lambda_{2,3}^B = u_0 \left(1 - \frac{(\gamma + 1)\rho^{\gamma+1}}{\beta - \rho^{\gamma+1}}\right)$$

3.8.4 Third System Model

For the third nonlinear system model given by (3.10), (3.32), and (3.33) we linearize the dynamics around some solution with a small perturbation. The result is given by (3.71), and the following two equations

$$\bar{v}_t + v_0 \bar{v}_x + u_0 \bar{v}_y + \rho_0 (V')^2 \bar{\rho}_x + \rho_0 (V')^2 \bar{\rho}_y = \frac{V'\bar{\rho} - \bar{v}}{\tau} \qquad (3.79)$$

$$\bar{u}_t + v_0 \bar{u}_x + u_0 \bar{u}_y + \rho_0 (U')^2 \bar{\rho}_x + \rho_0 (U')^2 \bar{\rho}_y = \frac{U'\bar{\rho} - \bar{u}}{\tau} \qquad (3.80)$$

In vector form (3.74), the linearized model is given by

$$\begin{bmatrix} \bar{\rho} \\ \bar{v} \\ \bar{u} \end{bmatrix}_t + \begin{bmatrix} v_0 & \rho_0 & 0 \\ \rho_0 (V')^2 & v_0 & 0 \\ \rho_0 (U')^2 & 0 & v_0 \end{bmatrix} \begin{bmatrix} \bar{\rho} \\ \bar{v} \\ \bar{u} \end{bmatrix}_x$$

$$+ \begin{bmatrix} u_0 & 0 & \rho_0 \\ \rho_0 (V')^2 & u_0 & 0 \\ \rho_0 (U')^2 & 0 & u_0 \end{bmatrix} \begin{bmatrix} \bar{\rho} \\ \bar{v} \\ \bar{u} \end{bmatrix}_y = \begin{bmatrix} 0 \\ (V'\bar{\rho} - \bar{v})/\tau \\ (U'\bar{\rho} - \bar{u})/\tau \end{bmatrix}.$$

3.8 Linearization

The system eigenvalues are found to be

$$\lambda_1 = v_0, \lambda_{2,3} = v_0 \pm \rho_0 V', \quad \text{and} \quad \lambda_1 = u_0, \lambda_{2,3} = u_0 \pm \rho_0 U',$$

which are the time-invariant version of the ones found in Sect. 3.6.3. This concludes the linearization section.

Chapter 4

Numerical Methods

4.1 Introduction

This chapter presents numerical schemes that approximate the solution of the hyperbolic, non-linear, time-varying, partial differential equations that represent crowd models developed in Chap. 3. Due to the hyperbolic nature of PDEs, discontinuous solutions like shocks and rarefaction waves can occur. Therefore, we use *shock capturing finite difference schemes* called Finite Volume Methods (FVM) [64, 67, 97]. These kind of schemes are capable of automatically choosing the correct weak solution, including shocks. This can be achieved because these methods are based on the integral form of the conservation laws which allow discontinuous solution, and not on the differential form where discontinuous solutions are not defined.

For the purpose of finding the numerical solution to the crowd models we use three first order accurate methods. The methods used are mainly, the Lax-Friedrichs scheme, First-Order Centered scheme (FORCE), and Roe's scheme. Higher order methods can be applied to obtain more accurate results but they are not considered here. The numerical solution can also be obtained by composite schemes, using first order Lax-Friedrichs method to smooth shocks in conjunction with a second order Lax-Wendroff method to increase accuracy [69], by using second or higher order schemes with flux limiters to smooth the shock waves instability (Roe's scheme can be modified to be a second order accurate). One family of schemes can not be applied on the models of Chap. 3, they are the flux vector splitting methods.

This is due to the homogeneity property $F(Q) = A(Q)Q$ which is not satisfied in our crowd models. The organization of this chapter is as follows, Sect. 4.2 presents the theory behind the FVM, followed by the description of three schemes.

4.2 Fundamentals of FVM

A finite volume method is based on dividing the space into grid cells or finite volumes and approximating the integral of the state Q over each volume. In each time step we update these values using approximations to the flux through the endpoints $[x_{i-1/2}, x_{i+1/2}]$ of the intervals as given by

$$Q_i^n \approx \frac{1}{\Delta x} \int_{x_{i-1/2}}^{x_{i+1/2}} q(x, t_n) \, dx \equiv \frac{1}{\Delta x} \int_{x_{i-1/2}}^{x_{i+1/2}} \tag{4.1}$$

where $\Delta x = x_{i+1/2} - x_{i-1/2}$ is the length of each cell on a uniform grid. By using a conservative numerical method we will guarantee that $\sum_{i=1}^{N} Q_i^n \Delta x$ over the entire space will change only due to fluxes at the boundaries. We will show this by recalling the integral conservation form (2.9) and rewrite it as

$$\frac{d}{dt} \int_{x_{i-1/2}}^{x_{i+1/2}} q(x, t) \, dx = f_{\text{in}}(q(x_{i-1/2}, t)) - f_{\text{out}}(q(x_{i+1/2}, t)). \tag{4.2}$$

For a given Q_i^n we would like to approximate Q_i^{n+1}, that is the volume average after $\Delta t = t_{n+1} - t_n$. To do so, we integrate (4.2) over time to obtain

$$\int_{x_{i-1/2}}^{x_{i+1/2}} q(x, t_{n+1}) dx - \int_{x_{i-1/2}}^{x_{i+1/2}} q(x, t_n) dx$$

$$= \int_{t_n}^{t_{n+1}} f_{\text{in}}(q(x_{i-1/2}, t)) - \int_{t_n}^{t_{n+1}} f_{\text{out}}(q(x_{i+1/2}, t)).$$

Dividing the above by Δx and rearranging we get

$$\frac{1}{\Delta x} \int_{x_{i-1/2}}^{x_{i+1/2}} q(x, t_{n+1}) \, dx = \frac{1}{\Delta x} \int_{x_{i-1/2}}^{x_{i+1/2}} q(x, t_n) \, dx$$

$$- \frac{1}{\Delta x} \left[\int_{t_n}^{t_{n+1}} f_{\text{out}}(q(x_{i+1/2}, t)) \, dt - \int_{t_n}^{t_{n+1}} f_{\text{in}}(q(x_{i-1/2}, t)) \, dt \right],$$

4.2 Fundamentals of FVM

and by comparing with (4.2), this translate to numerical method of the following form

$$Q_i^{n+1} = Q_i^n - \frac{\Delta t}{\Delta x}[F_{i+1/2}^n - F_{i-1/2}^n], \qquad (4.3)$$

where

$$F_{i+1/2}^n \approx \frac{1}{\Delta t} \int_{t_n}^{t_{n+1}} f(q(x_{i+1/2}, t)) \, dt.$$

The difference in finding the fluxes F at the boundary points of each cell is how these schemes vary (examples are in next section). Lets go back to the point of conservation, if we take $\sum_{i=I}^{J} Q_i^{n+1}\Delta x$ we will get

$$\sum_{i=I}^{J} Q_i^{n+1}\Delta x = \sum_{i=I}^{J} Q_i^n \Delta x - \frac{\Delta t}{\Delta x}[F_{J+1/2}^n - F_{I-1/2}^n]. \qquad (4.4)$$

that shows very clearly the conservative nature of this method. Here the changes to the hyperbolic PDE if any are due to the flow at the extreme boundary points only.

4.2.1 Formulation of 2-D Numerical Schemes

We are trying to solve the following problem

$$\left.\begin{array}{rl} PDEs & : \; Q_t + F(Q)_x + G(Q)_y = 0 \\ ICs & : \; Q(x,y,0) = Q^0 \\ B.Cs & : \; non-slip \; walls, \; and/or \; one \; exit \end{array}\right\} \qquad (4.5)$$

We start by identifying the notations involved in this process. The solution space (x, y, t) is split up into a uniform computational grid, where the grid spaces in the x and y directions are given by Δx, and Δy respectively as shown in Fig. 4.1, and in time direction by Δt. The positions of the ith and jth nodes in the x and y directions, and nth node in time direction (x_i, y_j, t^n) are given by $(i\Delta x, j\Delta y, n\Delta t)$.

There are various ways of approximating the spatial and time derivatives. Here we will concentrate on schemes that approximate the time derivative Q_t by a one-sided approximation given by

$$Q_t \approx \frac{Q^{n+1} - Q^n}{\Delta t}.$$

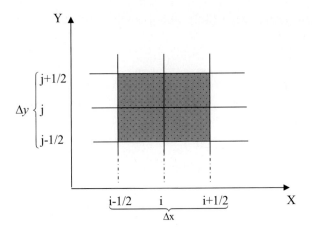

Fig. 4.1. Two-dimensional grid

By doing so, the schemes for 2-D space can be written in a general form as

$$Q_{i,j}^{n+1} = Q_{i,j}^n - \frac{\Delta t}{\Delta x}[F_{i+1/2,j}^n - F_{i-1/2,j}^n] - \frac{\Delta t}{\Delta x}[G_{i,j+1/2}^n - G_{i,j-1/2}^n]. \quad (4.6)$$

We use two approaches to solve the 2-D space problem. One is *fully discrete flux difference method* like Lax-Friedrichs scheme and obtain a solution according to (4.6), and another is *dimensional splitting* scheme like FORCE and Roe's. In the later method, we split the 2-D problem into a sequence of two 1-D problems as follows

$$\left.\begin{array}{rl} PDEs : & Q_t + F(Q)_x = 0 \\ ICs : & Q(x,y,0) = Q^0 \\ B.Cs : & \text{non-slip walls, and one exit} \end{array}\right\} \xrightarrow{\Delta t} Q^{n+\frac{1}{2}} \quad (4.7)$$

$$\left.\begin{array}{rl} PDEs : & Q_t + G(Q)_y = 0 \\ ICs : & Q^{n+\frac{1}{2}} \\ B.Cs : & \text{non-slip walls, and one exit} \end{array}\right\} \xrightarrow{\Delta t} Q^{n+1} \quad (4.8)$$

and use

$$Q_{i,j}^{n+\frac{1}{2}} = Q_{i,j}^n - \frac{\Delta t}{\Delta x}[F_{i+1/2,j}^n - F_{i-1/2,j}^n] \quad (4.9)$$

$$Q_{i,j}^{n+1} = Q_{i,j}^{n+\frac{1}{2}} - \frac{\Delta t}{\Delta x}[G_{i,j+1/2}^{n+\frac{1}{2}} - G_{i,j-1/2}^{n+\frac{1}{2}}]. \quad (4.10)$$

4.3 Numerical Schemes

4.3.1 Lax-Friedrichs Scheme

Lax-Friedrichs is a classical scheme with first order accuracy that solves (4.5) using (4.6) with numerical fluxes calculated at the interface. For 1-D and 2-D the scheme is given by

$$Q_i^{n+1} = \frac{1}{2}[Q_{i-1}^n + Q_{i+1}^n] - \frac{\Delta t}{2\Delta x}[F(Q_{i+1}^n) - F(Q_{i-1}^n)] \quad (4.11)$$

$$Q_{i,j}^{n+1} = \frac{1}{4}[Q_{i-1,j}^n + Q_{i+1,j}^n + Q_{i,j-1}^n + Q_{i,j+1}^n] - \frac{\Delta t}{2\Delta x}[F(Q_{i+1,j}^n)$$
$$- F(Q_{i-1,j}^n)] - \frac{\Delta t}{2\Delta x}[G(Q_{i,j+1}^n) - G(Q_{i,j-1}^n)]. \quad (4.12)$$

The scheme was selected because it is less expensive (takes less time computationally), and gives a good idea of the expected solution behavior. This method has numerical diffusion that damps the instabilities arising when shock solution is approximated. The method in 1-D is stable for $\nu = \Delta t/\Delta x \max_p |\lambda^p| \leq 1$ with λ as one of the system eigenvalues. For 2-D, we use *von Neumann Stability Analysis* [67] to get ν.

4.3.2 FORCE Scheme

The First Order Centered (FORCE) method is a first order accurate *total variation diminishing* (TVD) scheme that comes from the mean fluxes of the Lax-Friedrichs (4.11) and the Richtmyer second order accurate scheme [97]. The FORCE scheme solves numerically (4.5) by (4.7), and (4.8). The scheme computes a numerical flux by first defining an intermediate state $Q_{i-1}^{n+\frac{1}{2}}$, then the flux is computed based on this new intermediate state. Mathematically for 1-D problem this translates to

$$Q_{i+\frac{1}{2}}^{n+\frac{1}{2}} = \frac{1}{2}[Q_i^n + Q_{i+1}^n] - \frac{\Delta t}{2\Delta x}[F_{i+1}^n - F_i^n] \quad (4.13)$$

$$Q_{i-\frac{1}{2}}^{n+\frac{1}{2}} = \frac{1}{2}[Q_{i-1}^n + Q_i^n] - \frac{\Delta t}{2\Delta x}[F_i^n - F_{i-1}^n]. \quad (4.14)$$

then,

$$Q_i^{n+1} = \frac{1}{2}[Q_{i+\frac{1}{2}}^{n+\frac{1}{2}} + Q_{i-\frac{1}{2}}^{n+\frac{1}{2}}] - \frac{\Delta t}{2\Delta x}[F_{i+\frac{1}{2}}^{n+\frac{1}{2}} - F_{i-\frac{1}{2}}^{n+\frac{1}{2}}] \quad (4.15)$$

where
$$F_{i+\frac{1}{2}}^{n+\frac{1}{2}} = F(Q_{i+\frac{1}{2}}^{n+\frac{1}{2}}). \tag{4.16}$$

For the 2-D problem we use i,j instead of i as in (4.9) and (4.9) and use *dimensional splitting* method.

4.3.3 Roe's Scheme

The idea behind Roe's scheme is to take a non-linear PDE system in quasi-linear form
$$Q_t + A(Q)Q_x = 0. \tag{4.17}$$
and linearize locally by approximating the Jacobian matrix $A(Q)$ on an interval using Roe averages and in every time step repeat the process. The resulting system can then approximate speed found from the eigenvalues of the averaged Jacobian matrix $A(\tilde{Q})$ (see below). For the purpose of this section we define Q_R and Q_L that represent the states on the right and left in any direction from the flow boundary points $i \pm 1/2$. This will make the 2-D problem easier. First step in this process is to linearize locally, given by

$$F(Q) = F(Q_L) + A(\tilde{Q})[Q - Q_L] \tag{4.18}$$
$$F(Q) = F(Q_R) + A(\tilde{Q})[Q - Q_R] \tag{4.19}$$

then add the above two equations to get
$$F(Q_R) - F(Q_L) = A(\tilde{Q})[Q_R - Q_L] \tag{4.20}$$
where here we solve for the averages in $A(\tilde{Q})$, and substitute them in
$$F_{i+\frac{1}{2}}(Q_L, Q_R) = \frac{1}{2}[F(Q_L) + F(Q_R)] - \frac{1}{2}\tilde{R}|\tilde{\Lambda}|\tilde{R}^{-1}[Q_R - Q_L] \tag{4.21}$$
to obtain our approximated solution $A(\tilde{Q}) = \tilde{R}|\tilde{\Lambda}|\tilde{R}^{-1}$, where R is an eigenvector and Λ is diagonal matrix of the eigenvalues. The same thing is done for $G(Q)$.

For the first system crowd model the averages are found to be

$$\tilde{\rho} = \sqrt{\rho_L \rho_R} \tag{4.22}$$

$$\tilde{v} = \frac{v_L \sqrt{\rho_L} + v_R \sqrt{\rho_R}}{\sqrt{\rho_L} + \sqrt{\rho_R}} \tag{4.23}$$

$$\tilde{u} = \frac{u_L \sqrt{\rho_L} + u_R \sqrt{\rho_R}}{\sqrt{\rho_L} + \sqrt{\rho_R}}. \tag{4.24}$$

4.4 Simulation

This chapter presents the numerical solutions for the four models developed in Chap. 3 simulated by the numerical methods discussed in Chap. 4. We show the Roe scheme result for the first crowd system given by (3.18), while the Lax-Friedrichs and FORCE methods results are applied on the four 2-D models. Test problem were conducted on each model to produce a behavior close to what we would physically expect from real observation. The tests fall under two main types:

1. Gaussian density distribution with different initial velocities and changing the magnitude and sign of v_{f1} and v_{f2}.

2. Gaussian density distribution with one exit at a boundary.

We analysis and compare the results from all the models. First, we start with initial and boundary conditions.

4.4.1 Initial and Boundary Conditions

In the test problems we use gaussian distribution in 2-D space given by

$$\rho(x,y,0) = C \exp^{(-(x-a)^2 - (y-b)^2)}. \qquad (4.25)$$

where C is the maximum density value, and a & b determines the center of the density distribution. The free flow average speed (preferred speed) is assumed to be $v_{f1} = v_{f2} = 1.36\,\text{ms}^{-1}$ as many studies have used this value (it depends on service level concept), while density values vary from one study to another due to weather conditions, and location at the time of the study (winter, summer, men, children, Virginia, Moscow, etc.). Density is the number of pedestrians per unit area, and for the purpose of this simulation we set the maximum density $\rho_m = 1$.

Simulation is done on a square area with no obstacles that can divert the crowd flow, therefore, pedestrian are assumed to move freely within the boundaries. At the boundary non-slip conditions are enforced, except when we have an exit at some point (x_e, y_e) (i.e., large room, one exit, and without any obstacles). By non-slip we mean closed walls where no pedestrian (density) can pass through, but they can move tangent to the walls. We use ghost cells, which is basically an imaginary extra cells that works as an extra boundary layer on the outer sides of the grid to simulate the boundary effect as described in [67]. The exit is a free flow point, where we assume that once a pedestrian reach the exit cell, his exiting velocity is equal to the free flow velocity v_{f1} or v_{f2} (sucking point).

For the crowd to move toward the exit, we simply point them to it by forcing them to follow the desired direction using the velocity-density function $V(\rho)$, and this is done by

$$V = V(\rho)\cos\theta = V(\rho)\frac{x_e - x_i}{\sqrt{(x_e - x_i)^2 + (y_e - y_i)^2}} \qquad (4.26)$$

$$U = U(\rho)\sin\theta = U(\rho)\frac{y_e - y_i}{\sqrt{(x_e - x_i)^2 + (y_e - y_i)^2}}. \qquad (4.27)$$

4.4.2 Simulation Results

First model (one-equation model): We have two simulation results, the first show the model dynamic response to direction change as the velocity sign (direction) change to obtain the contours in Fig. 4.2. This show the model ability to have a controlled bi-direct- ional crowd flow and the response is consistent with the actual crowd behavior. When the space is not a constraint, pedestrians will move with free flow speed and diffuse in their given direction as time progresses. The scale of the diffuse in this simulation is bigger than the later figures due to simulation space and computational time. To distinguish this model dynamics from the others. We compare the result given in Fig. 4.3d with the Fourth model result in Fig. 4.17d. We notice that the inner contour in both plots are different, the one corresponding to the one-equation model is smaller. In addition, due to the lake of the relaxation term the direction change occur instantaneously by $\pm v_{f_{1,2}}$ for one-equation model. Therefor, pedestrians are moving faster as seen by reaching the bottom left corner in part (d).

4.4 Simulation

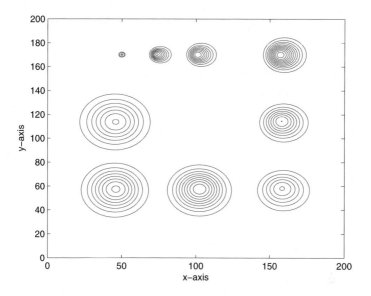

Fig. 4.2. Test 1 Contours of the density response for the one-equation model using Lax-Friedrichs method

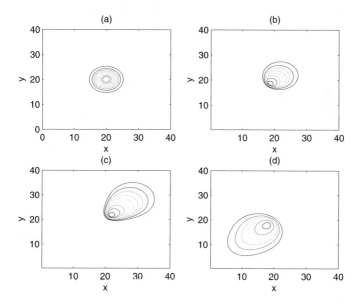

Fig. 4.3. Test 1 Contours of the density response to direction change at different time frames for the one-equation model using Lax-Friedrichs scheme

Second model (or first system model): Here we did several test to show the anticipation factor effect on the model. In Figs. 4.4 and 4.5 we changed the anticipation factor C_0 from 0.5 in the first to 1.1 in the second. The effect is clear from the figures part (d) plots. The first exceeded the maximum jam density, while the second its on target. In Fig. 4.6, and it corresponding contours in Fig. 4.7 show the effect of the shock wave formation as crowd move toward a corner in a closed area. In Fig. 4.8, we introduced an exit and the results show (a) how crowd move toward the exit, and (b) traffic jam at the exit point. In Figs. 4.9 and 4.10 we compare Roe's and FORCE schemes. Roe's scheme results are more concentrated than the FORCE scheme as seen from the center contour ring in part (d) in both figures.

Third model (or second system model): To compare this model with the previous one, we did the same last test as shown in Fig. 4.11. Although FORCE scheme is used, we got a response that is almost similar to the one in Fig. 4.9 for Roe's scheme. This shows the effect of the numerical schemes and their accuracy. In Fig. 4.12, we introduced an exit to simulate the second test response. We can see from the results how crowd flow toward the exit, and traffic jam is obvious at the exit point.

Fourth model (or the third system model): The density contours shown in Fig. 4.17 correspond to the density response when the initial free flow velocities are positive and equal (i.e., $v_{f1} = v_{f2}$). After some time the sign is changed to negative as seen from the system simulation in part (c). The results show that as time progresses, crowd (density) moves as expected toward the right upper corner, and then they change direction. At the same time the variation in the density concentration is changing as seen by plots (b) and (c) (diffusing in the direction of the flow). Figure 4.13, and its corresponding contours in Fig. 4.14 show a clear picture of the density flow. Here y-axis velocity is positive and x-axis is negative. The last test is to show the model response at the exit (see Figs. 4.15 and 4.16). The simulation predicts jam density at the exit which is verified by the observed behavior in an emergency evacuation near bottleneck areas.

4.4 Simulation

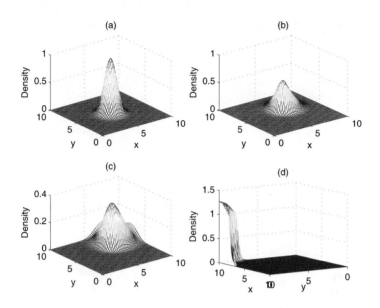

Fig. 4.4. Test 1 Density response at different time frame for the first system model with $C_0 = 0.5$ using Roe scheme

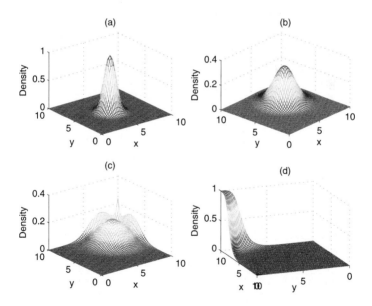

Fig. 4.5. Test 1 Density response at different time frame for the first system model with $C_0 = 1.1$ using Roe scheme

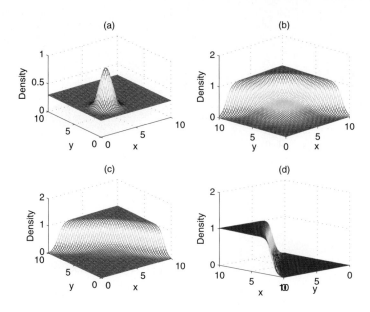

Fig. 4.6. Test 1 Density response at different time frame for the first system model with $C_0 = 1.1$ using Roe scheme

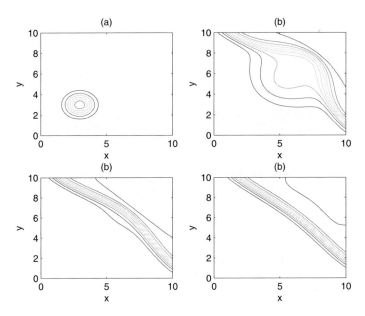

Fig. 4.7. Test 1 Contours of the density response at different time frame for the first system model with $C_0 = 1.1$ using Roe scheme

4.4 Simulation

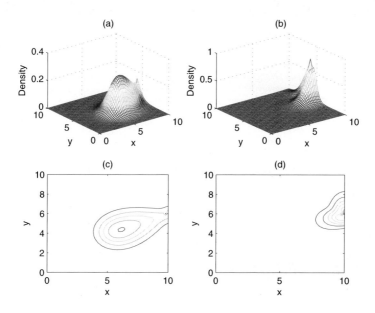

Fig. 4.8. Test 2 Density response at different time frame for the first system model with an exit. Simulated using Roe Scheme and $C_0 = 0.8$

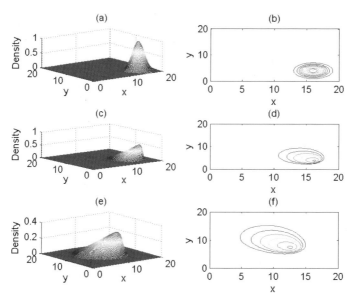

Fig. 4.9. First system model using Roe scheme, response at different time frame

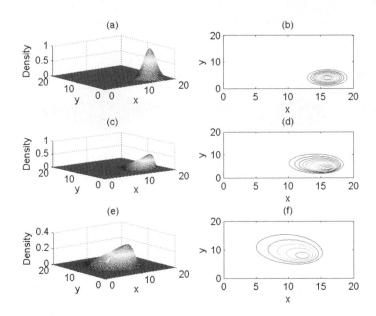

Fig. 4.10. First system model using FORCE scheme, response at different time frame

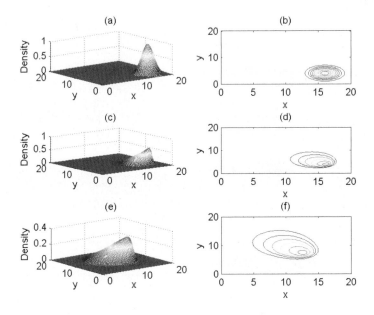

Fig. 4.11. Second system model using FORCE scheme, response at different time frame

4.4 Simulation

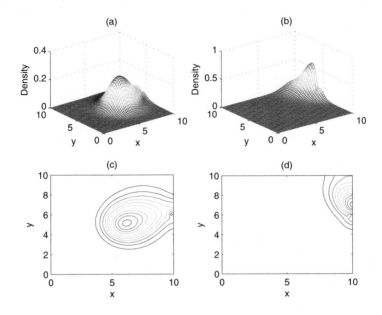

Fig. 4.12. Test 2 Density response at different time frame for the second system model with an exit. Simulated using Roe Scheme and $\gamma = 5$

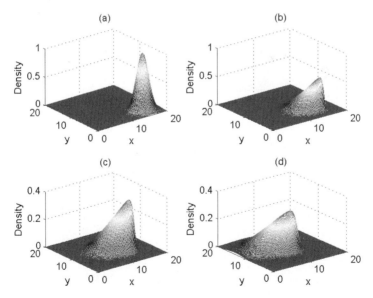

Fig. 4.13. Test 1 Density response to direction change at different time frames for the third system using FORCE scheme

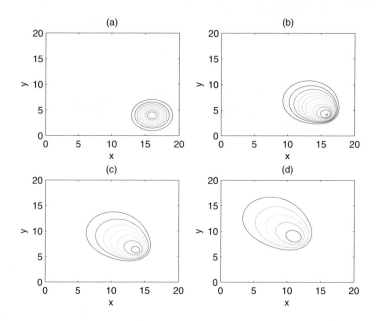

Fig. 4.14. Test 1 Contours of the density response to direction change at different time frames for the third system using FORCE scheme

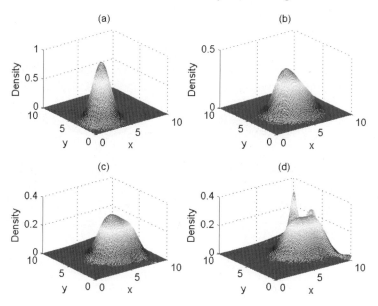

Fig. 4.15. Test 2 Density response directed toward an exit at different time frames for the third model using FORCE scheme

4.4 Simulation

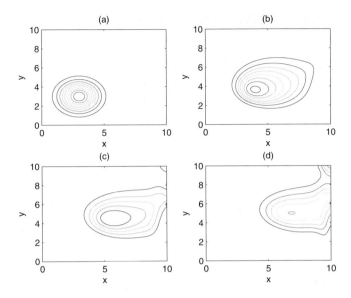

Fig. 4.16. Test 2 Contours of the density distribution directed toward an exit at different time frames for the third system using FORCE scheme

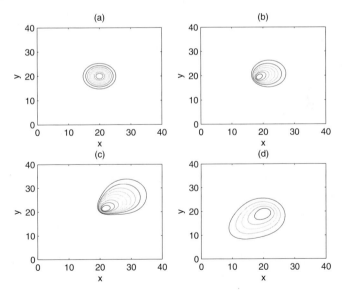

Fig. 4.17. Test 1 Contours of the density response to direction change at different time frames for the third system using Lax-Friedrichs scheme

4.5 Matlab Program Code

4.5.1 One-equation Model

```
%%%%%%%%%%%%%%%%%%%%%%%%%%%%%%%%%
% One-equation Model, FORCE scheme %
%%%%%%%%%%%%%%%%%%%%%%%%%%%%%%%%%
 D_t=0.04;   D_x=0.2;    D_y=0.2; max=40;
 time=450;  row_m=1; Vm=1.36; Vmm=1.36;
 x=[0:D_x:max];y=[0:D_y:max];
 N=length(x);
y1=0; for yy=0:D_y:max
    x1=0;y1=y1+1;
    for xx=0:D_x:max
     x1=x1+1;
     row_ic(y1,x1)=1*exp(-1/12*((xx-20)^2+(yy-20)^2));
    end
end row=row_ic;
%%%%%%%%%%%%%%%%
Nx=length(x); Ny=length(y); for t=1:time
    if t==180 ; Vm=-1.36;Vmm=-1.36;end
    for i=2:Ny
        for j=2:Nx
%Setting B.C, ghost cells, and speed = 0 at
% the boundary and calculate the fluxes
    if j==2;   %left B.C
    row(i,j-1)=row(i,j);
    f(i,j-1)=row(i,j-1)*-Vm*(1-row(i,j-1)./row_m);
    else
    f(i,j-1)=row(i,j-1)*Vm*(1-row(i,j-1)./row_m);
    end

    if j==Nx;  % Right B.C
    row(i,j+1)=row(i,j);
    f(i,j+1)=row(i,j+1)*-Vm*(1-row(i,j+1)./row_m);
    else
    f(i,j+1)=row(i,j+1)*Vm*(1-row(i,j+1)./row_m);
    end
```

4.5 Matlab Program Code

```matlab
        if i==2;  %bottom
        row(i-1,j)=row(i,j);
        g(i-1,j)=row(i-1,j)*-Vmm*(1-row(i-1,j)./row_m);
        else
        g(i-1,j)=row(i-1,j)*Vmm*(1-row(i-1,j)./row_m);
        end

        if i==Ny;   % Top
        row(i+1,j)=row(i,j);
        g(i+1,j)=row(i+1,j)*-Vmm*(1-row(i+1,j)./row_m);
        else
        g(i+1,j)=row(i+1,j)*Vmm*(1-row(i+1,j)./row_m);
        end
%    Updating the density "row"
        row_updated(i,j)=(row(i-1,j)+row(i+1,j)
                        +row(i,j-1)+...
        row(i,j+1))/4-(D_t/(2*D_x)*(f(i,j+1)-f(i,j-1))+...
        D_t/(2*D_y)*(g(i+1,j)-g(i-1,j)));
        end
        end
        row(2:Ny,2:Nx)=row_updated(2:Ny,2:Nx);
        r(1:Ny-1,1:Nx-1)=row(2:Ny,2:Nx);
end

%%%%%%%%%%%%%%%%%%%%%%%%%%%%%%%%%%%%%
% One-equation Model, FORCE scheme %
%%%%%%%%%%%%%%%%%%%%%%%%%%%%%%%%%%%%%
D_t=0.04;D_x=0.2;D_y=0.2;vf=1.36/2;vff=0*1.36;max=80;
Rm=1;max_t=2400;x=[0:D_x:max];y=[0:D_y:max];
N=length(x);
%   To find the I.C
y1=0;
for yy=0:D_y:max x1=0;y1=y1+1;
for xx=0:D_x:max x1=x1+1;
R_new(y1,x1)=1*exp(-1/2*((xx-10)^2+(yy-65)^2));
end
end
Rin=R_new;                       %I.C
```

```
for t=1:max_t           % time loop
% velocity direction
    if t==max_t/4;   vf=0*1.36;vff=-1.36/2;end
    if t==2*max_t/4; vf=-1.36/2;vff=0*1.36;end
    if t==3*max_t/4; vf=0*1.36;vff=1.36/2;end
% x and y loops
    i=2:N+1;
    j=2:N+1;
    R(j,i)=R_new(1:N,1:N);
    % Goust Cell B.C Down, Top, Right, Left
    R(1,i)=R(2,i); R(N+2,i)=R(N+1,i); R(j,1)=R(j,2);
    R(j,N+2)=R(j,N+1);
    for j=2:N+1
        for i=2:N+1
    [FL]=FORCE(R(j,i),R(j,i-1),vf,D_t,D_x,Rm);
    [FR]=FORCE(R(j,i+1),R(j,i),vf,D_t,D_x,Rm);
    R_04(j,i)=R(j,i)-D_t/(D_x)*[FR-FL];
        end
    end
    i=2:N+1;
    j=2:N+1;
    R_04(1,i)=R_04(2,i);R_04(N+2,i)=R_04(N+1,i);
    R_04(j,1)=R_04(j,2);R_04(j,N+2)=R_04(j,N+1);
    for j=2:N+1
        for i=2:N+1
    [GL]=FORCE(R_04(j,i),R_04(j-1,i),vff,D_t,D_y,Rm);
    [GR]=FORCE(R_04(j+1,i),R_04(j,i),vff,D_t,D_y,Rm);
    R_new(j-1,i-1)=R_04(j,i)-D_t/(D_y)*[GR-GL];
        end
    end
end

function [F]=FORCE(Rr,Rl,Vf,Dt,Dx,Rm);
    F_R=Rr*Vf*(1-Rr/Rm);
    F_L=Rl*Vf*(1-Rl/Rm);
    q_r=Rr;
    q_l=Rl;
    Q_RI=0.5*(q_l+q_r)+0.5*Dt/Dx*(F_L-F_R);
```

4.5 Matlab Program Code

```
    F_LF=0.5*(F_L+F_R)+0.5*Dx/Dt*(q_l-q_r);
    F_RI=Q_RI*Vf*(1-Q_RI/Rm);
    F=0.5*(F_LF+F_RI);
```

4.5.2 First System Model

```
%%%%%%%%%%%%%%%%%%%%%%%%%%%%%%%%%%%%%
% First System Model by Roe's scheme %
%%%%%%%%%%%%%%%%%%%%%%%%%%%%%%%%%%%%%
function Roe
clear all
%%%%%%%%%%%%%%%%%%%%%%%%%%%%%%%%%
%      Initial condition
%%%%%%%%%%%%%%%%%%%%%%%%%%%%%%%%%
D_t=0.04;D_x=0.2;D_y=0.2;tow=0.2;vf=1.36;
max=20;Rm=1;x=[0:D_x:max];y=[0:D_y:max];
N=length(x); y1=0;
for yy=0:D_y:max
 x1=0;y1=y1+1;
 for xx=0:D_x:max
  x1=x1+1;
  R_new(y1,x1)=1*exp(-1/4*((xx-16)^2+(yy-4)^2));
  U_new(y1,x1)=-vf*(1-R_new(y1,x1)/Rm);
  V_new(y1,x1)=vf*(1-R_new(y1,x1)/Rm);
 end
end
Rin=R_new;Uin=U_new;Vin=V_new;
%%%%%%%%%%%%%%%%%%%%%%%%%%%%%%%%%%%
for t=1:150
    i=2:N+1;
    j=2:N+1;
    R(j,i)=R_new(1:N,1:N);
    U(j,i)=U_new(1:N,1:N);
    V(j,i)=V_new(1:N,1:N);
    clear R_new U_new V_new Qold f_s FL FR GL GR...
        g_s Qold_04 Qold_34 Q_new Q_new_04 Q_new_34..
        R_04 U_04 V_04 R_34 U_34 V_34
    % Goust Cell B.C Down, Top, Right, Left
    R(1,i)=R(2,i);R(N+2,i)=R(N+1,i);R(j,1)=R(j,2);
```

```
R(j,N+2)=R(j,N+1);U(1,i)=U(2,i);U(N+2,i)=U(N+1,i);
U(j,1)=-U(j,2);U(j,N+2)=-U(j,N+1);V(1,i)=-V(2,i);
V(N+2,i)=-V(N+1,i);V(j,1)=V(j,2);V(j,N+2)=V(j,N+1);

for j=2:N+1
    for i=2:N+1

[FL]=Approx_Roe_PW(R(j,i),R(j,i-1),U(j,i),U(j,i-1),...
    V(j,i),V(j,i-1),0);
[FR]=Approx_Roe_PW(R(j,i+1),R(j,i),U(j,i+1),U(j,i),...
    V(j,i+1),V(j,i),0);

        Qold_04=[R(j,i);R(j,i)*U(j,i);R(j,i)*V(j,i)];
            q_new_04=Qold_04-D_t/(D_x)*[FR-FL];
            %UR=vf*(1-R(j,i)/1)*sign(U(j,i));
            r=q_new_04(1);
            u=q_new_04(2)/q_new_04(1);
            v=q_new_04(3)/q_new_04(1);
            UR=vf*(1-r/1)*sign(u);
            f_s=[0;r*(UR-u)/tow;0];%u
            Q_new_04=q_new_04+D_t*f_s;

            R_04(j,i)=Q_new_04(1);
            U_04(j,i)=Q_new_04(2)/Q_new_04(1);
            V_04(j,i)=Q_new_04(3)/Q_new_04(1);
        end
    end
    i=2:N+1;
    j=2:N+1;
    R_04(1,i)=R_04(2,i); R_04(N+2,i)=R_04(N+1,i);
    R_04(j,1)=R_04(j,2);R_04(j,N+2)=R_04(j,N+1);
    U_04(1,i)=U_04(2,i);U_04(N+2,i)=U_04(N+1,i);
    U_04(j,1)=-U_04(j,2);U_04(j,N+2)=-U_04(j,N+1);
    V_04(1,i)=-V_04(2,i);V_04(N+2,i)=-V_04(N+1,i);
    V_04(j,1)=V_04(j,2);V_04(j,N+2)=V_04(j,N+1);
    for j=2:N+1
        for i=2:N+1
    [GL]=Approx_Roe_PW(R_04(j,i),R_04(j-1,i),...
```

4.5 Matlab Program Code

```
            V_04(j,i),V_04(j-1,i),U_04(j,i),U_04(j-1,i),1);
            [GR]=Approx_Roe_PW(R_04(j+1,i),R_04(j,i),...
            V_04(j+1,i),V_04(j,i),U_04(j+1,i),U_04(j,i),1);

Qold_34=[R_04(j,i);R_04(j,i)*U_04(j,i);R_04(j,i)...
        *V_04(j,i)];
q_new_34=Qold_34-D_t/(D_x)*[GR-GL];
            r=q_new_34(1);
            u=q_new_34(2)/q_new_04(1);
            v=q_new_34(3)/q_new_34(1);
            VR=vf*(1-r/1)*sign(v);
            g_s=[0;0;r*(VR-v)/tow]; %v
            Q_new_34=q_new_34+D_t*g_s;

            R_new(j-1,i-1)=Q_new_34(1);
            U_new(j-1,i-1)=Q_new_34(2)/Q_new_34(1);
            V_new(j-1,i-1)=Q_new_34(3)/Q_new_34(1);
        end
    end
end

function [F]=Approx_Roe_PW(Rr,Rl,Ur,Ul,Vr,Vl,M);
c0=0.8;tow=0.2;
u=(Ur*(Rr)^(1/2)+Ul*(Rl)^(1/2))/((Rr)^(1/2)+(Rl)^(1/2));
v=(Vr*(Rr)^(1/2)+Vl*(Rl)^(1/2))/((Rr)^(1/2)+(Rl)^(1/2));

eig=[-c0+u  0   0;  0   u    0;  0  0  c0+u]; if M==0
    RR=[1   0    1; c0+u   0  -c0+u;  v   1     v];
    F_R=[Rr*Ur; Rr*Ur^2+c0^2*Rr; Rr*Ur*Vr];
    F_L=[Rl*Ul; Rl*Ul^2+c0^2*Rl; Rl*Ul*Vl];
    q_r=[Rr; Rr*Ur; Rr*Vr];
    q_l=[Rl; Rl*Ul; Rl*Vl];
else
    RR=[1   0   1; v   1    v; c0+u   0  -c0+u];
    F_R=[Rr*Ur; Rr*Ur*Vr; Rr*Ur^2+c0^2*Rr];
    F_L=[Rl*Ul; Rl*Ul*Vl; Rl*Ul^2+c0^2*Rl];
    q_r=[Rr; Rr*Vr; Rr*Ur];
    q_l=[Rl; Rl*Vl; Rl*Ul];
```

```
end
eps1=max([0, eig(1,1)-(c0+Ul), (c0+Ur)-eig(1,1)]);
Lamda_1=max(eps1,abs(eig(1,1)));
eps2=max([0,eig(2,2)-(Ul),(Ur)-eig(2,2)]);
Lamda_2=max(eps2,abs(eig(2,2)));
eps3=max([0,eig(3,3)-(-c0+Ul),(-c0+Ur)-eig(3,3)]);
Lamda_3=max(eps3,abs(eig(3,3))); eig_f=[Lamda_1 0...
 0; 0 Lamda_2 0;0 0 Lamda_3];

 Qa=RR*abs(eig_f)*inv(RR); F=0.5*[F_R +
F_L]-0.5*Qa*[q_r-q_l];

%%%%%%%%%%%%%%%%%%%%%%%%%%%%%%%%%%%%
% First System Model by FORCE scheme %
%%%%%%%%%%%%%%%%%%%%%%%%%%%%%%%%%%%%
function ForcedPW clear all
%%%%%%%%%%%%%%%%%%%%%%%%%%
%    Initial condition    %
%%%%%%%%%%%%%%%%%%%%%%%%%%
D_t=0.04;D_x=0.2;D_y=0.2;tow=0.2;vf=-1.36;vff=1.36;
max=20;Rm=1;c=0.8; x=[0:D_x:max];y=[0:D_y:max];
N=length(x); y1=0;
for yy=0:D_y:max
 x1=0;y1=y1+1;
 for xx=0:D_x:max
   x1=x1+1;
   R_new(y1,x1)=1*exp(-1/4*((xx-16)^2+(yy-4)^2));
   U_new(y1,x1)=-vf*(1-R_new(y1,x1)/Rm);
   V_new(y1,x1)=vff*(1-R_new(y1,x1)/Rm);
 end
end
Rin=R_new;Uin=U_new;Vin=V_new;
%%%%%%%%%%%%%%%%%%%%%%%%%%%%%%%%%%%%
for t=1:150
    i=2:N+1;
    j=2:N+1;
    R(j,i)=R_new(1:N,1:N);
    U(j,i)=U_new(1:N,1:N);
    V(j,i)=V_new(1:N,1:N);
```

4.5 Matlab Program Code

```
clear R_new U_new V_new Qold f_s FL FR GL GR...
  g_s Qold_04 Qold_34 Q_new Q_new_04 Q_new_34...
   R_04 U_04 V_04 R_34 U_34 V_34
% Goust Cell B.C Down, Top, Right, Left
R(1,i)=R(2,i); R(N+2,i)=R(N+1,i);
R(j,1)=R(j,2); R(j,N+2)=R(j,N+1);
U(1,i)=U(2,i);U(N+2,i)=U(N+1,i);
U(j,1)=-U(j,2); U(j,N+2)=-U(j,N+1);
V(1,i)=-V(2,i);V(N+2,i)=-V(N+1,i);
V(j,1)=V(j,2); V(j,N+2)=V(j,N+1);

for j=2:N+1
    for i=2:N+1
[FL]=FORCE(R(j,i),R(j,i-1),U(j,i),U(j,i-1),V(j,i),...
       V(j,i-1),D_t,D_x,c,0);
[FR]=FORCE(R(j,i+1),R(j,i),U(j,i+1),U(j,i),V(j,i+1),...
       V(j,i),D_t,D_x,c,0);

Qold_04=[R(j,i);R(j,i)*U(j,i);R(j,i)*V(j,i)];
q_new_04=Qold_04-D_t/(D_x)*[FR-FL];
Rr=q_new_04(1);
Uu=q_new_04(2)/(q_new_04(1));
UR=vf*(1-Rr/Rm);%*sign(Uu);
f_s=[0;Rr*(UR-Uu)/tow;0];
Q_new_04=q_new_04+D_t*f_s;
R_04(j,i)=Q_new_04(1);
U_04(j,i)=Q_new_04(2)/(Q_new_04(1));
V_04(j,i)=Q_new_04(3)/(Q_new_04(1));

end
end
    i=2:N+1;
    j=2:N+1;
    R_04(1,i)=R_04(2,i);    R_04(N+2,i)=R_04(N+1,i);
    R_04(j,1)=R_04(j,2);    R_04(j,N+2)=R_04(j,N+1);
    U_04(1,i)=U_04(2,i);    U_04(N+2,i)=U_04(N+1,i);
    U_04(j,1)=-U_04(j,2);   U_04(j,N+2)=-U_04(j,N+1);
    V_04(1,i)=-V_04(2,i);   V_04(N+2,i)=-V_04(N+1,i);
```

```
    V_04(j,1)=V_04(j,2);    V_04(j,N+2)=V_04(j,N+1);
    for j=2:N+1
        for i=2:N+1

[GL]=FORCE(R_04(j,i),R_04(j-1,i),U_04(j,i),
        U_04(j-1,i),...
    V_04(j,i),V_04(j-1,i),D_t,D_x,c,1);
[GR]=FORCE(R_04(j+1,i),R_04(j,i),U_04(j+1,i),
        U_04(j,i),...
    V_04(j+1,i),V_04(j,i),D_t,D_x,c,1);

Qold_34=[R_04(j,i);R_04(j,i)*U_04(j,i);R_04(j,i)...
        *V_04(j,i)];
 q_new_34=Qold_34-D_t/(D_x)*[GR-GL];

Rn=q_new_34(1);
 Vn=q_new_34(3)/(q_new_34(1));

VR=vff*(1-Rn/Rm);%*sign(Vn);

g_s=[0;0;Rn*(VR-Vn)/tow];
Q_new_34=q_new_34+D_t*g_s;

R_new(j-1,i-1)=Q_new_34(1);
U_new(j-1,i-1)=Q_new_34(2)/(Q_new_34(1));
V_new(j-1,i-1)=Q_new_34(3)/(Q_new_34(1));
        end
    end
end

function [F]=FORCE(Rr,Rl,Ur,Ul,Vr,Vl,Dt,Dx,c0,M);

if M==0
    F_R=[Rr*Ur; Rr*Ur^2+c0^2*Rr; Rr*Ur*Vr];
    F_L=[Rl*Ul; Rl*Ul^2+c0^2*Rl; Rl*Ul*Vl];
    q_r=[Rr; Rr*Ur; Rr*Vr];
    q_l=[Rl; Rl*Ul; Rl*Vl];
```

```
    F_LF=0.5*(F_L+F_R)+0.5*Dx/Dt*(q_l-q_r);
    Q_RI=0.5*(q_l+q_r)+0.5*Dt/Dx*(F_L-F_R);
    r=Q_RI(1);
    u=Q_RI(2)/(r);
    v=Q_RI(3)/(r);
    F_RI=[r*u; r*u^2+c0^2*r; r*u*v];
    F=0.5*(F_LF+F_RI);
else
    G_R=[Rr*Vr; Rr*Ur*Vr; Rr*Vr^2+c0^2*Rr];
    G_L=[Rl*Vl; Rl*Ul*Vl; Rl*Vl^2+c0^2*Rl];
    q_r=[Rr; Rr*Ur; Rr*Vr];
    q_l=[Rl; Rl*Ul; Rl*Vl];

    G_LF=0.5*(G_L+G_R)+0.5*Dx/Dt*(q_l-q_r);   %Lax
    Q_RI=0.5*(q_l+q_r)+0.5*Dt/Dx*(G_L-G_R);
    r=Q_RI(1);
    u=Q_RI(2)/(r);
    v=Q_RI(3)/(r);
    G_RI=[r*v; r*u*v; r*v^2+c0^2*r];    % Richm...
    F=0.5*(G_LF+G_RI);                  % Force
end
```

4.5.3 Second System Model

```
%%%%%%%%%%%%%%%%%%%%%%%%%%%%%%%%%%%%
% Second System Model by FORCE scheme %
%%%%%%%%%%%%%%%%%%%%%%%%%%%%%%%%%%%%
function ForcedAW clear all
%%%%%%%%%%%%%%%%%%%%%%%%%%
%    Initial condition %
%%%%%%%%%%%%%%%%%%%%%%%%%%
D_t=0.04;D_x=0.2;D_y=0.2;tow=0.2;vf=1.36;vff=1.36;
eps=0.000001;max=10;b=13;g=1;Rm=1;x=[0:D_x:max];
y=[0:D_y:max];N=length(x); y1=0;
for yy=0:D_y:max
 x1=0;y1=y1+1;
 for
  xx=0:D_x:max
   x1=x1+1;
```

```
    R_new(y1,x1)=1*exp(-1*((xx-5)^2+(yy-5)^2));
    U_new(y1,x1)=-vf*(1-R_new(y1,x1)/Rm);
    V_new(y1,x1)=vff*(1-R_new(y1,x1)/Rm);
      end
end
Rin=R_new;Uin=U_new;Vin=V_new;
%%%%%%%%%%%%%%%%%%%%%%%%%%%%%%%%%%%
for t=1:100
    i=2:N+1;
    j=2:N+1;
    R(j,i)=R_new(1:N,1:N);
    U(j,i)=U_new(1:N,1:N);
    V(j,i)=V_new(1:N,1:N);
clear R_new U_new V_new Qold f_s FL FR GL GR g_s...
  Qold_04 Qold_34 Q_new Q_new_04 Q_new_34 R_04 U_04...
    V_04 R_34 U_34 V_34
      % Goust Cell B.C Down, Top, Right, Left
      R(1,i)=R(2,i);R(N+2,i)=R(N+1,i);
      R(j,1)=R(j,2);R(j,N+2)=R(j,N+1);
      U(1,i)=U(2,i); U(N+2,i)=U(N+1,i);
      U(j,1)=-U(j,2);U(j,N+2)=-U(j,N+1);
      V(1,i)=-V(2,i);V(N+2,i)=-V(N+1,i);
      V(j,1)=V(j,2);V(j,N+2)=V(j,N+1);

      for j=2:N+1
          for i=2:N+1
[FL]=FORCE(R(j,i),R(j,i-1),U(j,i),U(j,i-1),V(j,i),...
    V(j,i-1),D_t,D_x,g,b,0,eps);
[FR]=FORCE(R(j,i+1),R(j,i),U(j,i+1),U(j,i),V(j,i+1),...
    V(j,i),D_t,D_x,g,b,0,eps);

P1=U(j,i)*R(j,i)^(g+1)/(b-R(j,i)^(g+1));
P2=V(j,i)*R(j,i)^(g+1)/(b-R(j,i)^(g+1));
Qold_04=[R(j,i);R(j,i)*(U(j,i)+P1);R(j,i)*(V(j,i)+P2)];
q_new_04=Qold_04-D_t/(D_x)*[FR-FL];

R_4(j,i)=q_new_04(1);
U_4(j,i)=q_new_04(2)*(b-q_new_04(1)^(g+1))/...
```

4.5 Matlab Program Code

```
            (b*q_new_04(1)+eps);
UR=vf*(1-R_4(j,i)/Rm);
f_s=[0;R_4(j,i)*(UR-U_4(j,i))/tow;0];
Q_new_04=q_new_04+D_t*f_s;

R_04(j,i)=Q_new_04(1);
U_04(j,i)=Q_new_04(2)*(b-R_04(j,i)^(g+1))/...
          (b*R_04(j,i)+eps);
V_04(j,i)=Q_new_04(3)*(b-R_04(j,i)^(g+1))/...
          (b*R_04(j,i)+eps);
        end
    end
    i=2:N+1;
    j=2:N+1;
    R_04(1,i)=R_04(2,i);    R_04(N+2,i)=R_04(N+1,i);
    R_04(j,1)=R_04(j,2);    R_04(j,N+2)=R_04(j,N+1);
    U_04(1,i)=U_04(2,i);    U_04(N+2,i)=U_04(N+1,i);
    U_04(j,1)=-U_04(j,2);   U_04(j,N+2)=-U_04(j,N+1);
    V_04(1,i)=-V_04(2,i);   V_04(N+2,i)=-V_04(N+1,i);
    V_04(j,1)=V_04(j,2);    V_04(j,N+2)=V_04(j,N+1);

    for j=2:N+1
        for i=2:N+1
[GL]=FORCE(R_04(j,i),R_04(j-1,i),U_04(j,i),
        U_04(j-1,i),...
      V_04(j,i),V_04(j-1,i),D_t,D_x,g,b,1,eps);
[GR]=FORCE(R_04(j+1,i),R_04(j,i),U_04(j+1,i),
        U_04(j,i),...
      V_04(j+1,i),V_04(j,i),D_t,D_x,g,b,1,eps);

pp1=U_04(j,i)*R_04(j,i)^(g+1)/(b-R_04(j,i)^(g+1));
pp2=V_04(j,i)*R_04(j,i)^(g+1)/(b-R_04(j,i)^(g+1));
Qold_34=[R_04(j,i);R_04(j,i)*(U_04(j,i)+pp1);R_04(j,i)...
        *(V_04(j,i)+pp2)];
q_new_34=Qold_34-D_t/(D_x)*[GR-GL];

R_n(j,i)=q_new_34(1);
V_n(j,i)=q_new_34(3)*(b-q_new_34(1)^(g+1))/...
```

```
            (b*q_new_34(1)+0.000001);
VR=vff*(1-R_n(j,i)/1);%*sign(V_n(j,i));
g_s=[0;0;R_n(j,i)*(VR-V_n(j,i))/tow];
Q_new_34=q_new_34+D_t*g_s;
R_new(j-1,i-1)=Q_new_34(1);
U_new(j-1,i-1)=Q_new_34(2)*(b-Q_new_34(1)^(g+1))/...
           (b*Q_new_34(1)+eps);
V_new(j-1,i-1)=Q_new_34(3)*(b-Q_new_34(1)^(g+1))/...
           (b*Q_new_34(1)+eps);
           end
     end
end

function [F]=FORCE(Rr,Rl,Ur,Ul,Vr,Vl,Dt,Dx,g,b,M,eps);
Pl_1=Rl^(g+1)*Ul/(b-Rl^(g+1));  Pr_1=Rr^(g+1)*Ur...
 /(b-Rr^(g+1));
Pl_2=Rl^(g+1)*Vl/(b-Rl^(g+1));  Pr_2=Rr^(g+1)*Vr...
 /(b-Rr^(g+1));

if M==0
    F_R=[Rr*Ur; Rr*Ur*(Ur+Pr_1); Rr*Ur*(Vr+Pr_2)];
    F_L=[Rl*Ul; Rl*Ul*(Ul+Pl_1); Rl*Ul*(Vl+Pl_2)];
    q_r=[Rr; Rr*(Ur+Pr_1); Rr*(Vr+Pr_2)];
    q_l=[Rl; Rl*(Ul+Pl_1); Rl*(Vl+Pl_2)];

    F_LF=0.5*(F_L+F_R)+0.5*Dx/Dt*(q_l-q_r);
    Q_RI=0.5*(q_l+q_r)+0.5*Dt/Dx*(F_L-F_R);
    r=Q_RI(1);
    u=Q_RI(2)*(b-r^(g+1))/(b*r+eps);
    v=Q_RI(3)*(b-r^(g+1))/(b*r+eps);
    p1=r^(g+1)*u/(b-r^(g+1));
    p2=r^(g+1)*v/(b-r^(g+1));
    F_RI=[r*u; r*u*(u+p1); r*u*(v+p2)];
    F=0.5*(F_LF+F_RI);
else
    G_R=[Rr*Vr; Rr*Vr*(Ur+Pr_1); Rr*Vr*(Vr+Pr_2)];
    G_L=[Rl*Vl; Rl*Vl*(Ul+Pl_1); Rl*Vl*(Vl+Pl_2)];
    q_r=[Rr; Rr*(Ur+Pr_1); Rr*(Vr+Pr_2)];
```

4.5 Matlab Program Code

```
        q_l=[Rl; Rl*(Ul+Pl_1); Rl*(Vl+Pl_2)];

        G_LF=0.5*(G_L+G_R)+0.5*Dx/Dt*(q_l-q_r);      %Lax
        Q_RI=0.5*(q_l+q_r)+0.5*Dt/Dx*(G_L-G_R);
        rr=Q_RI(1);
        uu=Q_RI(2)*(b-rr^(g+1))/(b*rr+eps);
        vv=Q_RI(3)*(b-rr^(g+1))/(b*rr+eps);
        pp1=rr^(g+1)*uu/(b-rr^(g+1));
        pp2=rr^(g+1)*vv/(b-rr^(g+1));
        G_RI=[rr*vv; rr*vv*(uu+pp1); rr*vv*(vv+pp2)];
        F=0.5*(G_LF+G_RI);                         % Force
end
```

4.5.4 Third System Model

```
%%%%%%%%%%%%%%%%%%%%%%%%%%%%%%%%%%%%%%%
% Third System Model by Lax scheme    %
%%%%%%%%%%%%%%%%%%%%%%%%%%%%%%%%%%%%%%%
function ForcedZhang
clear all
%%%%%%%%%%%%%%%%%%%%%%%%%%%
%    Initial condition   %
%%%%%%%%%%%%%%%%%%%%%%%%%%%
D_t=0.04;D_x=0.2;D_y=0.2;max=20;time=150;row_m=1;
Vm=1.36;Vmm=-1.36;x=[0:D_x:max];y=[0:D_y:max];
 N=length(x);
y1=0; for yy=0:D_y:max
    x1=0;y1=y1+1;
    for xx=0:D_x:max
    x1=x1+1;
R_new(y1,x1)=1*exp(-1/2*((xx-10)^2+(yy-10)^2));
V_new(y1,x1)=Vm*(1-R_new(y1,x1)/row_m);
U_new(y1,x1)=Vmm*(1-R_new(y1,x1)/row_m);
    end
end Ri=R_new;Ui=U_new;Vi=V_new;
%%%%%%%%%%%%%%%%%%%%%%%%%%%%%%%%%%%%%%%
for t=1:time
    R(2:N+1,2:N+1)=R_new(1:N,1:N);
    V(2:N+1,2:N+1)=V_new(1:N,1:N);
```

```
      U(2:N+1,2:N+1)=U_new(1:N,1:N);
      for i=2:N+1
          for j=2:N+1
% Setting B.C, ghost cells and speed=0 at the boundary
%   and calculating the fluxes
              if j==2;    %left B.C
                  R(i,j-1)=R(i,j);
                  V(i,j-1)=-V(i,j);
                  U(i,j-1)=U(i,j);
              end
              if j==N+1;  % Right B.C
                  R(i,j+1)=R(i,j);
                  V(i,j+1)=-V(i,j);
                  U(i,j+1)=U(i,j);
              end
              if i==2;    %bottom
                  R(i-1,j)=R(i,j);
                  V(i-1,j)=V(i,j);
                  U(i-1,j)=-U(i,j);
              end
              if i==N+1;  % Top
                  R(i+1,j)=R(i,j);
                  V(i+1,j)=V(i,j);
                  U(i+1,j)=-U(i,j);
              end

[Q_new]=FORCE(R(i,j-1),R(i,j+1),R(i-1,j),R(i+1,j),...
V(i,j-1),V(i,j+1),V(i-1,j),V(i+1,j),U(i,j-1),U(i,j+1),...
    U(i-1,j),U(i+1,j),D_t,D_y,D_x,Vm,Vmm,row_m);

              R_new(i-1,j-1)=Q_new(1);
              V_r_n=Vm*(1-Q_new(1)/row_m);
              U_r_n=Vmm*(1-Q_new(1)/row_m);
              V_new(i-1,j-1)=Q_new(2)/Q_new(1)+V_r_n;
              U_new(i-1,j-1)=Q_new(3)/Q_new(1)+U_r_n;
          end
      end
```

4.5 Matlab Program Code

```
end

function [q_new]=FORCE(RL,RR,RB,RU,VL,VR,VB,VU,UL,...
    UR,UB,UU,Dt,Dy,Dx,vf,vff,Rm);
            Vl=vf*(1-RL/Rm);
            Ul=vff*(1-RL/Rm);
            Vr=vf*(1-RR/Rm);
            Ur=vff*(1-RR/Rm);
            Vb=vf*(1-RB/Rm);
            Ub=vff*(1-RB/Rm);
            Vu=vf*(1-RU/Rm);
            Uu=vff*(1-RU/Rm);

            ql=[RL;RL*(VL-Vl);RL*(UL-Ul)];
            qr=[RR;RR*(VR-Vr);RR*(UR-Ur)];
            qb=[RB;RB*(VB-Vb);RB*(UB-Ub)];
            qu=[RU;RU*(VU-Vu);RU*(UU-Uu)];

            fl=[RL*VL;RL*VL*(VL-Vl);RL*VL*(UL-Ul)];
            fr=[RR*VR;RR*VR*(VR-Vr);RR*VR*(UR-Ur)];

            gb=[RB*UB;RB*UB*(VB-Vb);RB*UB*(UB-Ub)];
            gu=[RU*UU;RU*UU*(VU-Vu);RU*UU*(UU-Uu)];
%   Updatting the density "R"
            q_new=(ql+qr+qb+qu)./4-(Dt/(2*Dx))...
            *(fr-fl)+Dt/(2*Dy)*(gu-gb));
```

Chapter 5

Feedback Linearization (1-D Patches)

5.1 Introduction

In this chapter we present a controller design for the LWR one-equation model given in Chap. 2 using feedback linearization method. The usual approach for the control of PDEs is by controlling the linear or nonlinear ODEs that result from spatial discretization of the original PDEs. Known difficulties and disadvantages associated with this approach are, for example controllability and observability that should depends only on the locations of sensors and actuators, may also depend on the discretization points number, location, and method [85]. In addition, moving from infinite dimensional to finite dimensional systems may lead to conclusions about the stability of the open-loop and/or closed-loop system that are not generally correct. Therefore, significant amount of interest is aimed toward the development of a control design based on the distributed parameter systems which accounts for the nature of the PDEs [18] and references therein.

Here we are interested in designing a feedback controller to evacuate pedestrians from a 1-D area (e.g., corridor). The method of feedback linearization works in a way to cancel the non-linearities in the system and it is well-know in nonlinear control for ODEs [59], and it was introduced to quasi-linear hyperbolic PDEs in [18]. This chapter presents the feedback linearization control design for the LWR model.

This chapter is organized in the following manner: In Sect. 5.2, we give a mathematical background on the control design using state feedback linearization. Section 5.3 presents an application of the controller and discusses its closed-loop stability and simulation results. In Sect. 5.4, we extend this approach by dividing the 1-D area into n number of sections (also called patches).

5.2 Theory

5.2.1 Control Problem

The general quasi-linear 1-D PDE system can be written as

$$\frac{\partial Q}{\partial t} = A(Q)\frac{\partial Q}{\partial x} + f(Q) + g(Q)\bar{u},$$
$$y = h(Q), \qquad (5.1)$$
$$\text{B.C} : C_1\rho(a,t) + C_2\rho(b,t) = R(t)$$

where $Q(x,t)$ is the vector of state variables defined on $Q(x,t) \in H^2[(a,b),\Re]$. Using the patch idea (see Fig. 5.1) we set

$$\bar{u}(x,t) = \sum_{i=1}^{n} b^i(x) u^i(t) \qquad (5.2)$$

$$y^i(t) = \mathbf{C}^i \bar{y}(x,t) = \mathbf{C}^i h(Q) = \int_{x_i}^{x_{i+1}} c^i(x) h(Q(x,t))\, dx \qquad (5.3)$$

where $\mathbf{C}^i$ is an operator that depends on the desired performance given by an integral same as in the majority of the applications in

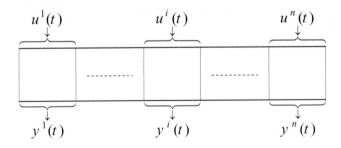

Fig. 5.1. Control specification in case of a 1-D problem

5.2 Theory

[18], $\mathbf{c}^i(x)$ and $\mathbf{b}^i(x)$ are known smooth functions on x assumed to be normalized and are given by

$$\sum_{i=1}^{n} \int_{x_i}^{x_{i+1}} b^i(x)\, dx = \sum_{i=1}^{n} \int_{x_i}^{x_{i+1}} c^i(x)\, dx = 1 \quad (5.4)$$

Using the above notation we rewrite the quasi-linear PDE to get

$$\begin{aligned}
\frac{\partial Q}{\partial t} &= A(Q)\frac{\partial Q}{\partial x} + f(Q) + g(Q)b(x)u, \\
y &= \mathbf{C}h(Q), \\
\text{B.C} \; &: \; C_1\rho(a,t) + C_2\rho(b,t) = R(t)
\end{aligned} \quad (5.5)$$

where $u = [u^1 \ldots u^l]^T$, and $y = [y^1 \ldots y^l]^T$.

5.2.2 Charactcristic Index

The concept of characteristic index σ is simply the smallest number of Lie derivatives (the derivative of a scalar field in the direction of a vector field) that establishes a direct relationship between the output y and the input u. The characteristic index can be found from

$$y = \mathbf{C}h(Q)$$

$$\frac{dy}{dt} = \mathbf{C}\left(\sum_{j=1}^{l}\frac{\partial Q_j}{\partial x}L_{a_j} + L_f\right)h(Q)$$

$$\vdots$$

$$\frac{d^\sigma y}{dt^\sigma} = \mathbf{C}\left(\sum_{j=1}^{l}\frac{\partial Q_j}{\partial x}L_{a_j} + L_f\right)^\sigma + CL_g\left(\sum_{j=1}^{l}\frac{\partial Q_j}{\partial x}L_{a_j} + L_f\right)^{\sigma-1}h(Q)b(x)u.$$

Once the term that includes the input u is nonzero we stop and proceed to the following step in control design. If $\sigma = \infty$, then u does not show up and this design method will not work.

5.2.3 State Feedback Control

Assuming all states are available for measurement, the controller we seek can be found from

$$\gamma_\sigma \frac{d^\sigma y}{dt^\sigma} + \ldots + \gamma_1 \frac{dy}{dt} + y = v \quad (5.6)$$

and for our system (5.5) it is given by

$$u = \left[\gamma_\sigma \mathbf{C} L_g \left(\sum_{j=1}^{n} \frac{\partial Q_j}{\partial x} L_{a_j} + L_f\right)^{\sigma-1} h(Q)b(x)\right]^{-1}$$

$$\times \left\{v - \mathbf{C}h(Q) - \sum_{v=1}^{\sigma} \gamma_v \mathbf{C} \left(\sum_{j=1}^{n} \frac{\partial Q_j}{\partial x} L_{a_j} + L_f\right)^v h(Q)\right\} \quad (5.7)$$

5.2.4 Closed-Loop Stability

Here we use the definition of closed-loop stability given in [18] for the system (5.5) under the controller (5.7). The controlled system is exponentially stable (i.e., the differential operator of the linearized closed-loop system generates an exponentially stable semigroup) if the following conditions are satisfied:

1. The roots of the equation

$$1 + \gamma_1 s + \ldots + \gamma_\sigma s^\sigma = 0$$

   lie in the open left hand of the complex plain.

2. The zeros dynamics of the system (5.5) are locally exponentially stable.

The first condition can be designed easily by pole placement, and the second condition is checked by setting the output $y = \mathbf{C}h(Q) = 0$. For more details see [18, 85].

5.3 Application to the LWR (One patch)

Our aim is to control the LWR model using one patch only. The one-equation PDE is given by

$$\begin{aligned}
\frac{\partial \rho}{\partial t} &= -\frac{\partial q}{\partial x} \\
&= -\frac{\partial \left(\rho v_f \left(1 - \frac{\rho}{\rho_m}\right)\right)}{\partial x} = -v_f \frac{\partial \left(1 - \frac{2\rho}{\rho_m}\right)}{\partial x} \\
&= -v_f \frac{\partial \tilde{q}(\rho)}{\partial x}
\end{aligned} \quad (5.8)$$

5.3 Application to the LWR (One patch)

where $u = v_f$ is the control variable, and the density ρ is the variable we want to control (for pedestrian evacuation, final $\rho = 0$). The area is $[0, 1] \in \Re$ and the boundary condition is closed from the left and open at the right end. Comparing this PDE to the general form (5.1) we find that $f(Q) = 0$, and $g(Q) = 0$. If we take $a = \frac{\partial \tilde{q}}{\partial \rho}$ the system can be represented by

$$\frac{\partial \rho}{\partial t} = -v_f \, a \, \frac{\partial \rho}{\partial x} \qquad (5.9)$$

where $a = \left(1 - \frac{2\rho}{\rho_m}\right)$. The choice of $\mathbf{c}^i(x) = \frac{1}{x_{i+1} - x_i}$ agrees with (5.4) and gives

$$y = \mathbf{C}h(\rho) = \int_0^1 \frac{1}{1-0} \rho(x,t) \, dx = \int_0^1 \rho(x,t) \, dx \qquad (5.10)$$

where the output operator $\mathbf{C}$ is one patch only, and $b(x) = 1$. Next we proceed with the control design as follows:

1. For the characteristic index if we just take the derivative of the output one time v_f will show as:

$$\begin{aligned} y &= \mathbf{C}h(Q) = \mathbf{C}\rho(x,t) \\ \dot{y} &= \mathbf{C}\frac{\partial \rho}{\partial \rho}\dot{\rho} = \mathbf{C}\frac{\partial \rho}{\partial \rho}\frac{\partial \rho}{\partial t} \\ &= -\mathbf{C}v_f \, a \, \frac{\partial \rho}{\partial x} = \mathbf{C} - v_f \frac{\partial \tilde{q}}{\partial x} \\ &= -v_f \int_0^1 \frac{\partial \tilde{q}}{\partial x} \, dx \\ &= -v_f(\tilde{q}(1) - \tilde{q}(0)). \end{aligned}$$

Hence, the characteristic index is equal to one. The manipulated input v_f is outside of the integral because it depends on time only. The integral and the partial derivative operator cancel out and we are left with the $\tilde{q}$ evaluated at the boundary points.

2. The state feedback controller comes from (5.6) which in our case reduces to $\dot{y} + K\,y = 0$, where $v = 0$ and $K = \frac{1}{\gamma_1}$. Thus

the controller is given by

$$v_f = u = \frac{K \int_0^1 \rho(x,t) \mathrm{d}x}{(\tilde{q}(1) - \tilde{q}(0))}. \qquad (5.11)$$

3. The closed-loop system can be found by substituting (5.11) in (5.8) to get

$$\frac{\partial \rho(x,t)}{\partial t} = -\left\{ \frac{K \int_0^1 \rho(x,t) \mathrm{d}x}{(\tilde{q}(1) - \tilde{q}(0))} \right\} \times \frac{\partial \tilde{q}(\rho)}{\partial x}. \qquad (5.12)$$

For the stability of the closed-loop system, and according to the conditions given in Sect. 5.2.3 we first choose $\gamma > 0$ or $K > 0$ because $1 + \gamma_1 s = 0 \implies s = -\frac{1}{\gamma_1}$ is in the left half of

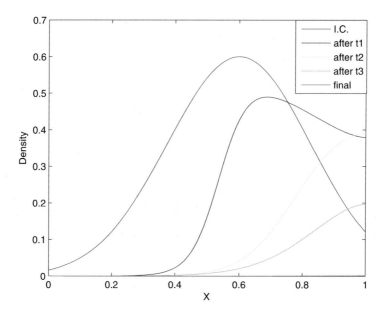

Fig. 5.2. Density response for one patch controller. Initial density maximum value is 0.6, $\Delta t = 0.002$ sec and $\Delta x = 0.01$, gain K=1, and total time ≈ 7 sec. to reach zero density

the complex plain. For the zero dynamics we set $y = 0$, and this can happen if and only if $\rho = 0$ because the density is a non-negative value (($y = 0$) $\iff$ ($\rho = 0$)) and we get

$$\frac{\partial \rho(x,t)}{\partial t} = -\frac{K \overbrace{\int_0^1 \rho(x,t)\, dx}^{=0}}{(\tilde{q}(1) - \tilde{q}(0))} \times \frac{\partial \tilde{q}}{\partial x}.$$

$$\frac{\partial \rho}{\partial t} = 0$$

Hence, zero dynamics are stable. Therefore the closed-loop system is stable.

The control action response for a corridor with exit at the right hand side is shown in Fig. 5.2. After some finite time, evacuation is completed and the density is zero. This controller attempt to control the flow around $y = \int_0^1 \rho(x,t)\, dx$ every time which is shown in the figure by the different time responses.

5.4 Application to the LWR ($n=5$ patches)

Here we repeat the design steps 1 to 4 shown in the previous section with five patches instead of one. Therefore, we have i=1,...5 and $b^i u^i = u^i$, where b^i are taken as unity distribution functions and

$$u(t) = \begin{cases} u^1(t), & [0.0, 0.2] \\ u^2(t), & [0.2, 0.4] \\ u^3(t), & [0.4, 0.6] \\ u^4(t), & [0.6, 0.8] \\ u^5(t), & [0.8, 1.0] \end{cases} \tag{5.13}$$

The desired performance is to control the outputs

$$y(t) = \begin{cases} y^1(t) = \int_{0.0}^{0.2} \rho(x,t) \mathrm{d}x \\ y^2(t) = \int_{0.2}^{0.4} \rho(x,t) \mathrm{d}x \\ y^3(t) = \int_{0.4}^{0.6} \rho(x,t) \mathrm{d}x \\ y^4(t) = \int_{0.6}^{0.8} \rho(x,t) \mathrm{d}x \\ y^5(t) = \int_{0.8}^{1} \rho(x,t) \mathrm{d}x \end{cases} \tag{5.14}$$

So the output is

$$y^i(t) = \int_{x_i}^{x^{i+1}} \mathbf{c}^i(x)\rho(x,t)\,dx \qquad (5.15)$$

where $\mathbf{c}^i(x) = \frac{1}{x_{i+1}-x_i}$.

1. The characteristic index is equal to one as shown by item 1 in Sect. 5.3.
2. The state feedback controller is given by

$$u^i = \frac{K \int_{x_i}^{x^{i+1}} \frac{\rho(x,t)}{x_{i+1}-x_i}\,dx}{(\tilde{q}(x_{i+1}) - \tilde{q}(x_i))}. \qquad (5.16)$$

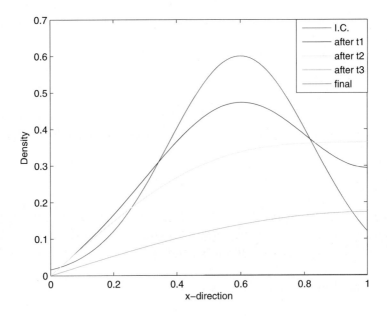

Fig. 5.3. Density response for five patch controller. Initial density maximum value is 0.6, $\Delta t = 0.000005$ sec and $\Delta x = 0.01$, gain K=1, and total time ≈ 0.2 sec. to reach zero density

3. Closed-loop system is given by

$$\frac{\partial \rho(x,t)}{\partial t} = -\left[\sum_{i=1}^{5}\left(H(x-x_i) - H(x-x_{i+1})\right)u^i\right] \times \frac{\partial \tilde{q}(\rho)}{\partial x}. \tag{5.17}$$

4. For closed-loop stability, condition one is satisfied if $K > 0$ as in the one patch case. For the zero dynamics we set $y = 0$, and this happens if and only if in each patch $\rho = 0$. Substitute $y = 0$ in the closed-loop in 3 we obtain $\frac{\partial \rho(x,t)}{\partial t} = 0$. Hence the closed-loop system is stable.

The control action response for a corridor with exit at the right hand side is shown in Fig. 5.3. After some finite time, evacuation is completed and the density is zero. The response here is different than the one for the single patch as seen by the responses approaching zero.

5.5 Matlab Program Code

5.5.1 One-patch Control

```
%%%%%%%%%%%%%%%%%%%%%%%%%%%%%%%%%%%
%   Control: One patch matlab Code    %
%%%%%%%%%%%%%%%%%%%%%%%%%%%%%%%%%%%
n=100;L=1; d_x=L/n;d_t=0.002; K=1.0; rm=1; h=d_x;
% n: number of points, L: length, K: gain,
% rm: jam density.
x=[0:d_x:1]; N=length(x);   % Corridor length
x1=0;                    % Start of the initial condition
for xx=0:d_x:1 x1=x1+1; Ro(x1)=0.6*exp(-10*(xx-0.6)^2);
end                      % End of the initial condition
r(1,1:n+1)=Ro(1:n+1);

for t=1:3500;     % Total time = t* d_t
% We first find the integral using trapezoid rule
% for every time step
    sum=0;
    for c=1:n
```

```
            sum=sum+(r(t,c)+r(t,c+1));
        end
% Next we compute the PDE
    for j=2:n+1
        if j==2;r(t,j-1)=0*r(t,j);end
        if j==n+1 ; r(t,j+1)=1*r(t,j);end
        q(t,j+1)=r(t,j+1)*(1-r(t,j+1)./rm);
        q(t,j-1)=r(t,j-1)*(1-r(t,j-1)./rm);
        r(t+1,j)=0.5*(r(t,j+1)+r(t,j-1))...
            -(d_t/d_x)/2*(K*h/2*sum)*(q(t,j+1)...
            -q(t,j-1))/((r(t,n+1).*(1-r(t,n+1)./rm))...
            -(r(t,2).*(1-r(t,2)./rm)));
    end
end
```

5.5.2 Five-patch Control

```
%%%%%%%%%%%%%%%%%%%%%%%%%%%%%%%%%
%  Control: 5 patches matlab Code  %
%%%%%%%%%%%%%%%%%%%%%%%%%%%%%%%%%
n=100;L=1; d_x=L/n;d_t=0.000005; h=d_x; K=1; rm=1;
% you can change d_t, gain K, and jam density rm.
x=[0:d_x:1]; N=length(x);  % Corridor length
x1=0;              % Start of the initial condition
for xx=0:d_x:1
x1=x1+1;
Ro(x1)=0.6*exp(-10*(xx-0.6)^2);
    end            % End of the initial condition

r(1,1:n+1)=Ro(1:n+1);

for t=1:20000;           % Total time = t* d_t
% We first find the integral using trapezoid rule for
% every time step for the five patches.
    sum_1=0;sum_2=0;sum_3=0;sum_4=0;sum_5=0;
    for c=1:n/5
        sum_1=sum_1+h/2*(r(t,c)+r(t,c+1))*1/0.2;
    end
    for c=n/5+1:n*2/5
```

5.5 Matlab Program Code

```
            sum_2=sum_2+h/2*(r(t,c)+r(t,c+1))*1/0.2;
        end
        for c=n*2/5+1:n*3/5
            sum_3=sum_3+h/2*(r(t,c)+r(t,c+1))*1/0.2;
        end
        for c=n*3/5+1:n*4/5
            sum_4=sum_4+h/2*(r(t,c)+r(t,c+1))*1/0.2;
        end
        for c=n*4/5+1:n
            sum_5=sum_5+h/2*(r(t,c)+r(t,c+1))*1/0.2;
        end
% Next we compute the PDE
        for j=2:n+1
            % no inflow from the left B.C.
            if j==2;r(t,j-1)=0*r(t,j);end
            % outflow from the right B.C.
            if j==n+1 ; r(t,j+1)=1*r(t,j);end
            % calculate the fluxes
            q(t,j+1)=r(t,j+1)*(1-r(t,j+1)./rm);
            q(t,j-1)=r(t,j-1)*(1-r(t,j-1)./rm);
% Next we update each patch
            if j<=n/5; sum=sum_1;
                q_R=r(t,n/5).*(1-r(t,n/5)./rm);
                q_L=r(t,2).*(1-r(t,2)./rm);

    r(t+1,j)=0.5*(r(t,j+1)+r(t,j-1))-(d_t/d_x)/2*(K*sum)...
                *(q(t,j+1)-q(t,j-1))/(q_R-q_L);
            end

            if j>n/5 & j<=2*n/5; sum=sum_2;
                q_L=r(t,n/5+1).*(1-r(t,n/5+1)./rm);
                q_R=r(t,2*n/5).*(1-r(t,2*n/5)./rm);
    r(t+1,j)=0.5*(r(t,j+1)+r(t,j-1))-(d_t/d_x)/2*(K*sum)...
                *(q(t,j+1)-q(t,j-1))/(q_R-q_L);
            end

            if j>2*n/5 & j<=3*n/5; sum=sum_3;
                q_L=r(t,2*n/5+1).*(1-r(t,2*n/5+1)./rm);
```

```
            q_R=r(t,3*n/5).*(1-r(t,3*n/5)./rm);
r(t+1,j)=0.5*(r(t,j+1)+r(t,j-1))-(d_t/d_x)/2*(K*sum)...
            *(q(t,j+1)-q(t,j-1))/(q_R-q_L);
        end

        if j>3*n/5 & j<=4*n/5; sum=sum_4;
            q_L=r(t,3*n/5+1).*(1-r(t,3*n/5+1)./rm);
            q_R=r(t,4*n/5).*(1-r(t,4*n/5)./rm);
r(t+1,j)=0.5*(r(t,j+1)+r(t,j-1))-(d_t/d_x)/2*(K*sum)...
            *(q(t,j+1)-q(t,j-1))/(q_R-q_L);
        end

        if j>4*n/5; sum=sum_5;
            q_L=r(t,4*n/5+1).*(1-r(t,4*n/5+1)./rm);
            q_R=r(t,n+1).*(1-r(t,n+1)./rm);
r(t+1,j)=0.5*(r(t,j+1)+r(t,j-1))-(d_t/d_x)/2*(K*sum)...
            *(q(t,j+1)-q(t,j-1))/(q_R-q_L);
        end
    end
end
```

Chapter 6

Intelligent Evacuation Systems

In this chapter we discuss the new concept of intelligent evacuation system or IES. This new technology will make a major change in evacuation strategies. Control, communication and computing technologies will be combined into IES system that can increase crowd safety and management without changing the physical structure of the facility. This chapter outlines the key features of medium area IES and shows how core crowd decisions may improve. We also propose a basic IES control system architecture based on the one given by [98], and we discuss the information technology issues related to its implementation.

6.1 Introduction

Congestion conditions during an emergency evacuation poses a great challenge for any evacuation management system. Congestion occur when the demand for travel exceeds the runway (corridor, stairs, highways), and bottleneck (exits, intersections) capacity. Hence a sound approach for reducing congestions will involve a mix of policies affecting demand and capacity depending on local circumstances and priorities. These policies include buildings, parking lots, stadiums, subway stations, airports, highways, and reducing demand by mass transit, carpooling on highways. This chapter discusses the potential of another strategy option called Intelligent Evacuation System

(IES). In this strategy, we claim that a proper combination of control, communication and computing technologies, placed on corridors, exits, TVs, internet, cars, and highways can assist people's decisions in ways that will increase evacuation safety and management without changing the physical structure of a facility or roads.

The intelligent evacuation system we propose here should take the following aspects into consideration

- Function: The range of functions associated with the evacuation system to be automated, and the degree of automations.

- Architecture: The breakdown of these functions into various control tasks, and the delegation of these tasks.

- Design: The division of intelligence between evacuees and facility, and how the use of technologies can realize this architecture.

- Effectiveness: The cost effectiveness and benefits of different evacuation systems.

Other aspects of this strategy relate to the timing of the system development, which concerns the accommodation of new functions in new design. This chapter is concerned with the above four points only.

Section 6.2 presents a framework for describing IES functions and their relation to key evacuee decision for different evacuation scenarios. Different IES proposals can be compared according to the degree of influence they exert on those decisions. These proposals may range from systems that are relatively simple from control viewpoint, since they merely provide information or advice to the evacuee, to more automated systems using open loop and feedback control strategies. In an open loop scheme the traffic will flow in a predetermined way which would be assessed based on the initial condition of the flow. Since the evacuation of people is a very dynamic situation hence a feedback solution is obviously better suited. We argue that an evacuation system that use feedback control strategy can achieve much better evacuation time reduction and management. The evacuation strategies will be optimally calculated on a real-time basis using advanced concepts of the feedback control theory.

Section 6.3 outlines two evacuation scenarios of evacuating people, train, and cars from a train substation, and an airport. For city

6.2 IPES Functions

Table 6.1. Evacuee decisions and IES functions

Stage	Evacuee decision	IES aim	IES job	Approach
Pre-evac	Evacuation plan and methodology	Efficient use of resource	Demand shift	1, 2, 4
In-evac	Route choice	Optimize travel time	Route guidance and flow control	1, 2, 3
	Path planning	Smooth flow	Congestion control	1, 2, 3
	Regulation	Increase flow, safety	Proper spacing, steering, etc.	1, 3
Post-evac	Parking, Shelters	Add safety	Efficient use	1, 2, 3, 4

evacuation we refer the reader to [98]. We discuss these scenarios in terms of the evacuee decisions listed in Table 6.1. Designing such systems poses a challenging problem for control.

Section 6.4 proposes a four-layer hierarchical control architecture which splits this problem into more manageable units. Starting at the top, the layers are called the network, link, planning and regulation layers. The regulation layer's task is to execute feedback laws for acceleration, direction change, and stop. The planning layer's task is to coordinate the movement of neighboring evacuee/vehicles and supervise the regulation layer during the execution of feedback laws. The link layer is responsible for the control of aggregated traffic flow over some area of the facility. Finally, the network layer assigns routes to evacuees/vehicles and controls admission into exits.

Section 6.5 discusses the control architecture for evacuation scenarios described in Sect. 6.3. Section 6.6 discusses the information technology issues.

Section 6.7 gives a brief discussion on feedback control and modeling of the crowd dynamics.

6.2 IPES Functions

Here we present five decisions made by an evacuee in the course of an emergency evacuation for small size area (includes floor, building, and parking lot), mid size area (subway station, airport), and large size area (includes city, county, and state). Table 6.1 gives these decisions and points out how they can be modified to increase safety and management of an evacuee. We divide the decisions into three stages, pre-evac, in-evac, and post-evac. Corresponding to these decisions, IES objectives and tasks are given, which if carried out accordingly, would realize the desired objectives.

We can add more decisions to this decision list, but in our view, the five decisions given in the table are the most important to IES plan and the rest do not affect it significantly. Also including more decisions to the list will increase its size considerably. In addition, people will be more comfortable with a system that can take care of these non-core issues. For instant, an automated exit doors can eliminate the need to stop, find the keys, and open these doors. Doors here could be building exit doors, subway or train doors, parking lot gates, highway toll, etc.

The last column in the table gives possible approaches to carry out the IES tasks, and these are

1. Supplying information (road condition, exit locations, etc.)

2. Giving advice.

3. Controlling the decisions.

4. Changing choices based on cost related issues (the use of a car, carpooling, toll, etc.)

These approaches increase the evacuation system difficulty, as we move from providing information to giving advice to applying control (can be either open loop or feedback) as represented by

$$\text{Information} \rightarrow \text{Advice} \rightarrow \text{Control}$$

However, at the same time evacuee decisions are becoming more automated as we go from providing information to applying control.

In the next section we see how the functions listed in Table 6.1 work for two different evacuation scenarios. Generally speaking, scenarios can be divided into three broad categories: small area, mid-size area, and large area. The small area scenarios cover the cases of a floor, building, and a parking lot. The mid-size area scenarios which we discuss next covers a subway station and an airport. The large area scenarios deal with evacuating people/cars from towns, cities etc.

6.3 IES Functions for Evacuation Scenarios

Here we will argue that a controlled evacuation system will provide better safety and management results than the information or advice evacuation strategies. To do that, we take a close look at two scenarios and discuss the effect of Table 6.1 on improving the evacuation process.

6.3.1 Subway Station

First scenario is a one level subway station given in Fig. 6.1. In a pre-evac phase, we will discuses how information, advice and cost can affect the evacuation generation and model choice, and how the IES system makes them more efficient. An information based evacuation system can consist of sensors connected to an alarm, and lights signs showing train doors and exit locations. An alarm will start and evacuees' decision will be to take the train, walk or run to safety (exit the station), or maybe ask for information. In an advice system, trained human resources can provide the decision by asking the evacuees to move slowly, run toward an exit, or take a shelter and so on. To make this phase more efficient a controlled process can provide the decision based on open loop or feedback controller. Possible

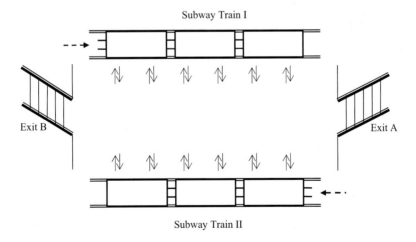

Fig. 6.1. Subway station

methods are voice commands, automatic doors, or light matrix that shows slow light changing for walking and fast for running. These methods are different in open loop from feedback based on sensing other information like jam at one of the exits, possibility of trains colliding (from network information), biological threat [80], etc.

The in-evac phase, the route choice decision for an information based system can be following signs or other evacuees toward what is supposed to be safety. However, this could guide to congestions at bottleneck locations (exit doors, and stairs) and increase evacuation time [49, 50]. Information that evacuees obtain from visual, and interactions with other evacuees/obstacles that is categorized under *behavior observations* of the traffic flow will influence their maneuver, and regulation decisions. For the advice system, route choice and path planning can be provided. Nevertheless, since advice is given based on visual area conditions, same bottleneck problem could still occur, even if advisors (trained human resources) are communication with each other. For example, they have no control on exit doors, changes to trains route, etc. To improve this phase, control design can choose the route with minimum time to reach an exit or safety based on current location and flow control. It also provides the evacuee with a plan to reach the exit by preventing congestion near bottleneck locations. In addition, to increase safety and management of evacuees, maneuver and regulation decisions can be used to coordinate exits (open and close), train doors, etc., and flow control (e.g., variation in moving velocities and direction). Finally, the post-evacuation phase is concerned with providing help by notifying emergency services, guide them to the disaster area and guiding injured evacuee to their parking area to get the fastest help possible.

6.3.2 Airport

An airport is considered a mid size area. Example of an airport schematic is given in Fig. 6.2. It consists of a terminal, airplanes, parking lots, building services, and vehicle routs. The aim here is to discuss this scenario and how the functions in Table 6.1 improve the evacuation process. Before we do so, we note that the plan is to evacuate all pedestrians in the affected area. If this area is a floor, building, or a parking lot we can create scenarios for them as well, but this is not considered here.

6.3 IES Functions for Evacuation Scenarios

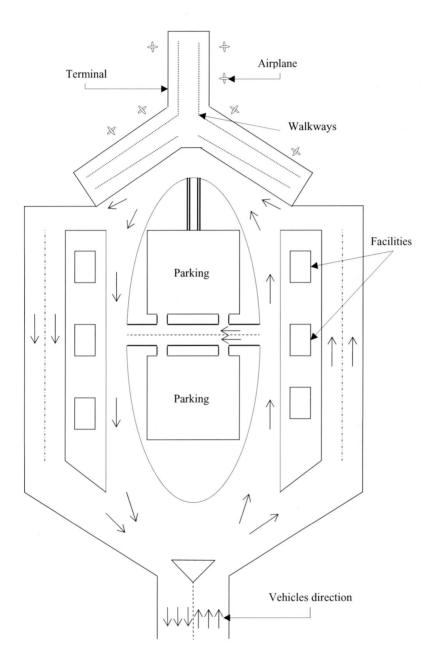

Fig. 6.2. Airport traffic flow

In pre-evac, a simple evacuation system can consist of sensors in a building connected to an alarm, and lights signs showing exit locations. When an alarm starts, evacuee's decision will be to ask for information, walk or run to safety following the exit signs that will lead them to exit the building. Next they go to the parking lot to their vehicles, or any other transportation available to drive them out of the airport area (disaster area). In an advice system, trained human resources can guide the evacuees to use certain routs, like taking refuge in a shelter and so on. But both systems can not deal with mass evacuation where congestion and jam will occur in buildings, parking lots, and on the roads connecting these facilities. To improve this process, an open loop or feedback controller can be used to assign an evacuation route, and synchronize the number of evacuees and vehicles flow into the roads. By doing so, smooth flow could be achieved. Possible methods are commands asking the controller of each node (facility) to evacuate a certain number of evacuees per unit time, automatic road and air traffic control, etc. These methods are different in open-loop from feedback based on sensing information like congestions, road blocks, number of airplanes landing and leaving from the airport, etc.

The in-evac phase, the route choice decision for a simple system can be self or following other evacuees that will eventually lead to congestions at bottleneck locations and increase evacuation time. For the advice system, route choice and path planning is based on visual and limited communication between system advisors (security personals, police). In addition, their control on roads, and airplanes traffic is insufficient (many airplane accidents happen in normal situation). To improve this phase, control design can choose the route with minimum time to reach an exit or safety based on current location and flow control of the different system nodes. Maneuver and regulation decisions can increase safety and management of evacuees by controlling velocity and direction of evacuees/vehicles. Finally, the post-evac phase is concerned with identifying an appropriate shelter, notifying emergency services and lead them to the disaster area, and guide injured evacuee to safe areas to get the fastest help possible.

6.4 Four-Layer System Architecture

With the above mentioned scenarios as background we now discuss how to formulate the problem of designing the control system which carries the tasks of IES. In this section we propose a control system architecture similar to one proposed in [98]. Here we will represent the functions given in Table 6.1 by a control system architecture whose block diagram representation is given by Fig. 6.3. From top, the layers are called network, link, planning and regulation.

There is a single network layer. The controller task in this layer is to assign an escape route for the evacuees in the system, and an entrance route for the emergency services. From abstract point of view, this controller must determine a routing table with entries of the form

$$(\text{Current evacuee location, exit}) \rightarrow (\text{Evacuee route})$$

This table is updated based on the area affected in a way to optimize evacuee flow. The updates would be based on the information about the aggregate state of the flow of people. To find the entries for this table, we make use of controllers that identify the route associated with optimum time. This can be done either by feedback or open-loop controller. A feedback assignment of routes would correspond to assessing the current traffic and incident situation, then deciding dynamically about the route assignment. In an open-loop scheme

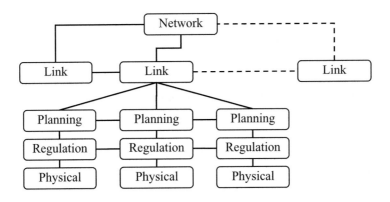

Fig. 6.3. Four layer architecture

the traffic will flow in a predetermined way which would be assessed based on the initial condition of the flow. Since the evacuation of people is a very dynamic situation, hence a feedback solution is obviously better suited. We suggest the use of central computer for the calculations of the entries, since on board calculation is difficult, because it requires equipments such as computing devices, sensing information, and communication to be available with every evacuee.

In the IES, the disaster area is divided into various sections with a link layer controller assigned to each section. Each link layer assigns target speed and path to all the evacuees in such a way that (a) evacuee follow the assigned rout toward the exit, (b) congestion in and between the links is avoided. Thus, if there is a jam condition at a front link, the control link layer must adopt and change its speed command (slow down or stop). The calculations are updated based on the response in density distribution and speed changes in each section. The updates depends on the link area size and they are made more frequently in this layer to assist the situation. The link layer controller design is based on the macroscopic models for crowd dynamics given in Chap. 3.

There is a planning layer and a regulation layer controller associated with each evacuee. The planning layer has a task of path planning. It generates a plan which closely follows the path or route assigned to evacuees by the link layer. Once a plan is generated the regulation layer's task is to execute a trajectory for an evacuee target speed which conforms with the plan.

The planning layer is assumed to have information at all times of the evacuees position, its path and target speed as assigned by link layer. It then decides what actions to take in order to evolve a plan which should stay as close as possible to the proposed path. In addition, it exchanges this information with the planning layers of neighboring evacuees to coordinate the action. Then, requests the regulation layer to implement the feedback control law to accomplish that action.

The regulation layer implements the necessary feedback law pertaining to the request from its planning layer. It informs the planning layer of completion of the action which then plans the next action. The feedback laws compute the value for the velocity vector actuators based on the sensor inputs of the evacuees present state. The physical layer is a macroscopic model of the evacuee density represented by partial differential equations. This layer has actuator inputs and sensor outputs.

The layers we have presented are in a general form. They can be applied on different physical scenarios and different choices of control authority based on the size of the problem.

6.5 Four-Layer System: Scenarios

In this section we discuss the control system architecture described in previous section for our scenarios.

6.5.1 Subway Station

Here we describe the four layer control system architecture for the subway station:

1. Network layer: This layer assigns route to the evacuees in the station that could be toward an exit or taking the train to the next or final stop. Feedback control is used to identify the route associated with optimum time. A feedback assignment of routes would correspond to assessing the current state of the situation (danger, evacuee density and location).

2. Link layer: Here links are section of the waiting area, train cabins, stairs, or a train in a subway network. The controller in this layer assigns a path to each evacuee such that the flow is smooth from all links. It also assigns a target speed for each link to ensure smooth flow and avoid congestion.

3. Planning layer: This layer decides the actions to take by evacuee in different links in order to stay as close as possible to the assigned path. It should coordinates the maneuvers of neighboring evacuees or trains such that there is no conflict between the evacuees or collision between trains as well as to ensure the safety of motion.

4. Regulation layer: Here the various control laws based on the control inputs given by the planning layer are implemented using the hardware in the physical layer.

6.5.2 Airport

In the airport scenario, the four layer control system architecture are given by the following:

1. Network layer: In this layer the feedback controller assigns routes to the evacuees that achieve optimum time based on their location, which might be airplane door, terminal exits, parking lot exit, on road or air traffic. A feedback assignment of routes would correspond to assessing the current conditions of the system (danger, road conditions, evacuee density and location).

2. Link layer: In this scenario links are roads, buildings, parking lots, airplanes. The controller in this layer assigns a path to each evacuee such that the flow is smooth from all links. It also assigns a target speed for each link to ensure smooth flow and avoid congestion.

3. Planning layer: This layer trays to keep evacuees as close as possible to their assigned path. Coordinates the maneuvers of neighboring evacuees, airplanes, and cars such that there is no conflict between the evacuees and to ensure the safety of motion.

4. Regulation layer: Here the various control laws based on the control inputs given by the planning layer are implemented using the hardware in the physical layer.

6.6 IT Issues and Requirements

From recent disasters (e.g., Hurricane Katrina), communication systems used by early responders agencies, such as, police, firefighters, costal guard, homeland security, and others lead to a slower and uncoordinated response. This problem is due to the lack of common standards (interoperability) between the different communications systems. When the lack of compatibility coupled with the loss of infrastructure, the needed resources are not dispatched where they are most wanted. For the control architecture given by Fig. 6.3 we suggest the use of a communication architecture that deals with the different communication issues in case of an evacuation.

Network topology deals with how the network is physically arranged, physically constructed, and electrically connected. It also have a set of rules that control communications (protocols). These issues need to be address for the IES system proposed. We require that information collecting and sharing within each layer and with other layers should be robust, and the different standards must be able to communicate with each other correctly.

To physically construct a network (i.e., the way a medium connect the network nodes) there are two ways. The wired media and its technologies, and second is the wireless topology. For the IES system the wireless topology can be used as a backup medium due to bad weather conditions associated with evacuations (like hurricanes, storms, etc.). Electrical topology would be the communication path between nodes and how they can share information and resources, i.e., the way the wires or cables are actually connected. One solution could be the use of a hybrid network, which is a combination of the three basic star, bus, and ring networks. This form provides redundancy of paths that will enable other path choices in case of some link failure. To have a valuable network, it should be

reliable error free, error detection system

available 24*7, from any point, continuous stable power source

perform provide stable bandwidth under varying conditions, such as fixed, or dynamic bandwidth, bandwidth on demand, and brusty traffic

secure physical, Virtual, data.

6.7 Feedback Control and Dynamic Modeling

The control laws executed by the regulation layers of IES will be based on the feedback control theory. It will interpret and synthesize the data collected from sensors (hazard detection sensors and video cameras) to formulate feedback control command to be delivered to the entrapped people through actuators (speakers, TV monitors, and lighted matrix displays). The system will continuously monitor, on a real-time basis, how the evacuation evolves including the development of congestion at bottleneck points, to re-direct people to other less congested locations if it is deemed necessary.

The optimal evacuation strategy will be established based on the most advanced mathematical concepts of the feedback control theory of distributed systems. To do control design in a microscopic model environment, we need to keep following each evacuee's position, its interactions with other evacuees, and give a targeted control input to each one. Obviously this increases the complexity of the system as the number of evacuees increase and it is practically impossible to control each evacuee. Therefore, the IES uses macroscopic dynamic models where the evacuees are treated as a continuum and the affected area can be divided to sections and the control commands are locally implemented.

The development of the preliminary theoretical framework for the feedback control design has been already done with controllers being designed for a simple evacuation scheme. This preliminary work is expected to provide a sound theoretical framework for the solution of the more complex evacuation problem. The feedback control design for distributed dynamics can be done either in distributed setting (pde control) or in lumped setting (ode control). The distributed feedback controllers have been developed in Chap. 5 for 1-D space and in [99, 100] for one and 2-D spaces respectively.

Chapter 7

Discretized Feedback Control

7.1 Introduction

Pedestrian evacuation is a multi disciplinary problem that poses various concerns for design of infrastructure, pedestrian behavior, strategy etc. Considerable effort has been put in understanding the pedestrian behavior in normal and panic situations. Hoogendoorn and Bovy [50] have developed a normative theory that proposes a rational optimization based basis for the tactical decisions involved in route choice and activity scheduling of pedestrians. In their theory they account for the uncertainties involved in such decision making and the fact that pedestrians can choose from an infinite number of paths. The walking behavior on which this tactical decision is based is presented by Hoogendoorn in [49]. Hill [47] has proposed that directness is an important aspect in route selection as it is done unconsciously. Other simulation based works on pedestrian route choice selection can be found in [12, 31]. Helbing et al. [41] have done simulations using a pedestrian behavior model and assessed the pedestrian behavior in panic situations. The design of infrastructure has to facilitate the evacuation by providing adequate space and exits. The design of infrastructure for evacuation has been addressed in various codes from various regulating agencies that specify certain minimum dimensions for various architectural elements like corridors, doors etc. and placement of exits, for example the life safety code handbook by NFPA [20]. Cellular automata formulations for simulation of pedestrian movement are also a widely used tool [15, 60, 61, 62] and [89]. A continuum theory for the pedestrian movement which is different from a typical fluids theory modeling pedestrians has been proposed by Hughes [51]. As far as strategy for evacuation is concerned it can be dealt with at the network level and link level. At the network level the main concern is that of assigning the various links for the evacuating population. This problem has been dealt with in the literature from both a theoretical and a practical perspective [5, 16, 21] and [72]. Control of vehicular traffic by sensing the state of the system is addressed formally in the dynamic traffic assignment literature [53, 54, 55, 56] and [57]. Barrett et al. [7] have done dynamic traffic assignment for hurricane related evacuations. In pedestrian evacuation smooth evacuation of the people without jamming is a major concern at the link level. This issue is addressed in the following sections.

In this chapter the focus will be on developing the controls for evacuating pedestrians from a corridor. The motion of the pedestrians is assumed to be along the length of the corridor and appropriate equations governing the motion of the pedestrians are proposed. Since the purpose of developing the equations governing the pedestrian evacuation motion is to develop controls for it hence the equations are purposely kept to be simplistic. The modeling of the pedestrian motion is done at the macroscopic level wherein the pedestrian flow is assumed to be a continuum. The continuity partial differential equation is used as the governing equation. The corridor is then divided into a finite number of sections, n for the purpose of discretizing the partial differential equation into n, *ode's* to develop the control. Each *ode* and control variable corresponds to a section of the corridor. Since the pedestrian flow is treated as a continuum hence the pedestrian densities in each section are assumed to be the state variables. The following steps roughly outline the procedure adopted to model the system and develop the control:

1. Divide the corridor into a finite number, n, of sections and discretize the governing continuity partial differential equation into n ordinary differential equations. Each equation corresponds to a section of the corridor. A schematic of the corridor and its division into sections is shown in Fig. 7.1.

7.2 Pedestrian Flow Modeling

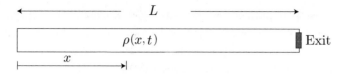

(a) One-dimensional Corridor

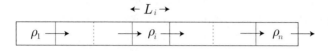

(b) Disretized model of the Corridor

Fig. 7.1. Schematic of a 1-D corridor

Corresponding to each section will be an input discharge an output discharge and a pedestrian density.

2. Assume a model relating the velocity of the pedestrian flow, v to the pedestrian density, ρ which has a control parameter associated with it. Use this relation to obtain a relation for the discharge or through-put of pedestrians and substitute in the above developed ordinary differential equations. Thus we have a ordinary differential equation in terms of the states and the control parameter.

3. Using the control parameter in the assumed v–ρ relationship develop a control based on certian criteria to control the ordinary differential equations developed in step 2.

7.2 Pedestrian Flow Modeling

As stated before continuity partial differential equation is used as the governing equation to model pedestrian evacuation flow. It relates the pedestrian density to the discharge. In 1-D this equation has the following form:

$$\frac{\partial \rho}{\partial t} = -\frac{\partial q}{\partial x}, \qquad (7.1)$$

In (7.1) ρ is the pedestrian density and q is the discharge. The time is denoted by t and x is the distance along the length of the corridor from the fixed end. For the purpose of developing the control the corridor is divided into a finite number n of sections. This is done so that the above partial differential equation (*pde*) can be discretized into n ordinary differential equations (*ode*'s). This is done by assuming no spatial variation of the pedestrian density in a section and integrating (7.1) over the length of the section. The following set of *ode*'s are obtained by doing so:

$$\dot{\rho}_1 = -\frac{q_1}{L_1}$$

$$\dot{\rho}_i = -\frac{q_i - q_{i-1}}{L_i}, \quad 2 \leq i \leq n, \qquad (7.2)$$

In (7.2) ρ_i denotes the density in the ith section and q_i denotes the output discharge of the ith section. L_i is the length of the section. This completes the 1st step in the procedure described above. Next we proceed with assuming the v–ρ model to obtain an expression for the discharge q in the above equations in terms of the densities and the control variables. A well established model in traffic theory is the *Greenshield's* model which assumes a linearly decreasing relationship between the flow velocity v and the pedestrian density ρ. The *Greenshield's* model can be expressed by the following equation below:

$$v = v_f(1 - \frac{\rho}{\rho_m}), \quad 0 \leq \rho \leq \rho_m, \qquad (7.3)$$

The parameter v_f in the above equation is the free flow velocity i.e., the velocity at $\rho = 0$. The parameter ρ_m is the jam density i.e., the density at which $v = 0$. The velocity v in (7.3) is the spatially averaged pedestrian velocity. Hence this velocity can be used to compute the discharge given by the following relation:

$$q = \rho v. \qquad (7.4)$$

Substituting (7.4) in (7.2) we get the following *ode*'s:

$$\dot{\rho}_1 = -\frac{\rho_1(1 - \frac{\rho_1}{\rho_m})v_{f_1}}{L_1},$$

$$\dot{\rho}_i = -\frac{\rho_i(1-\frac{\rho_i}{\rho_m})v_{f_i} - \rho_{i-1}(1-\frac{\rho_{i-1}}{\rho_m})v_{f_{i-1}}}{L_i}. \tag{7.5}$$

These equations are in the standard state-space format. We can normalize the above (7.5) by dividing all the equations by ρ_m. This results in the following set of equations:

$$\dot{\tilde{\rho}}_1 = -\tilde{\rho}_1(1-\tilde{\rho}_1)b_1\tilde{v}_{f_1}$$

$$\dot{\tilde{\rho}}_i = \tilde{\rho}_{i-1}(1-\tilde{\rho}_{i-1})b_i\tilde{v}_{f_{i-1}} - \tilde{\rho}_i(1-\tilde{\rho}_i)b_i\tilde{v}_{f_i} \tag{7.6}$$

where

$$\tilde{\rho}_i = \frac{\rho_i}{\rho_m}, \quad \tilde{v}_{f_i} = \frac{v_{f_i}}{L}, \quad b_i = \frac{L}{L_i}.$$

Henceforth we drop the "tilde" on the normalized quantities and hence (7.6) become:

$$\dot{\rho}_1 = -\rho_1(1-\rho_1)b_1 v_{f_1}$$

$$\dot{\rho}_i = \rho_{i-1}(1-\rho_{i-1})b_i v_{f_{i-1}} - \rho_i(1-\rho_i)b_i v_{f_i} \tag{7.7}$$

From now onwards (7.7) will be used to describe the dynamics of the pedestrian flow. The constraints in the *Greenshield's* model (7.3) translate into the following constraints for the state variables ρ_i in the normalized state equation (7.7):

$$0 \leq \rho_i \leq 1 \tag{7.8}$$

7.3 Feedback Linearization of State Equations

Linear systems are well understood from a controls and dynamics point of view. Observe that the state equations (7.7) are non-linear equations. The objective of this section is to choose the controls that will cancel the non-linearities in the above state equations. In doing so we will obtain a set of linear state equations that will hold for the entire state-space and make our analysis considerably easier. A simple control which can achieve this is:

$$v_{f_i} = \frac{k}{1-\rho_i}. \tag{7.9}$$

In (7.9) $k = $ const. is the control gain. If we substitute the control in (7.9) in (7.7) we will obtain the following equations:

$$\dot{\rho}_1 = -kb_1\rho_1$$

$$\dot{\rho}_i = kb_i\rho_{i-1} - kb_i\rho_i. \tag{7.10}$$

Observe that the non-linearities in (7.7) are not present in (7.10) owing to our specific choice of control. Thus using this control we are able to convert the non-linear equation (7.7) to linear equations (7.10) in the state variables ρ_i. Also we can see in (7.9) that the control v_{f_i} depends on the state variable ρ_i. Thus its a feedback control. Thus this is a feedback control that linearizes the system of governing differential equations. Hence it is a feedback linearizing control. Feedback linearization is a technique widely discussed in the standard texts on non-linear control [59, 76, 90]. In the next section we discuss the stability of the system under this control.

7.3.1 Stability Under Feedback Linearizing Control

Equation (7.9) can be written in the matrix form as:

$$\dot{\rho} = -k\mathbf{B}\rho. \tag{7.11}$$

In this equation $k > 0$ and $\mathbf{B}$ is a $n \times n$ matrix which has the following form:

$$\mathbf{B} = \begin{bmatrix} b_1 & & & & & \\ -b_2 & b_2 & & & & \\ 0 & -b_3 & b_3 & & & \\ 0 & & -b_4 & b_4 & & \\ \cdot & & & & & \\ \cdot & & & & & \\ \cdot & \cdot & \cdot & \cdot & -b_n & b_n \end{bmatrix} \tag{7.12}$$

The stability issues for a linear system like (7.11) are dealt with in standard texts on linear systems specifically [88]. The system of (7.11) is stable if the matrix $-k(\mathbf{B}^\mathbf{T} + \mathbf{B}) < 0$ (i.e. is negative

7.3 Feedback Linearization of State Equations

definate). Since $k > 0$ this is equivalent to $(\mathbf{B^T + B}) > 0$ (positive definate). This is equivalent to requiring that $\mathbf{D} = \frac{\mathbf{B^T + B}}{2} > 0$. The form which D assumes for a general n dimensional system is:

$$\mathbf{D} = \begin{bmatrix} b_1 & -\frac{b_2}{2} & & & & & \\ -\frac{b_2}{2} & b_2 & -\frac{b_3}{2} & & & & \\ 0 & -\frac{b_3}{2} & b_3 & -\frac{b_4}{2} & & & \\ 0 & & -\frac{b_4}{2} & b_4 & \cdot & & \\ \cdot & & & \cdot & \cdot & & \\ \cdot & & & & \cdot & & -\frac{b_n}{2} \\ \cdot & \cdot & \cdot & \cdot & & -\frac{b_n}{2} & b_n \end{bmatrix} \quad (7.13)$$

For $\mathbf{D} > 0$ we need the determinants D_i of the i dimensional matrices $\mathbf{D}_i$ having the 1st i elements along the diagonal of $\mathbf{D}$ along its diagonal. If we observe the structure of $\mathbf{D}$ then we can conclude the following reccurence relation for D_i:

$$D_1 = b_1$$

$$D_2 = b_1 b_2 - \frac{b_2^2}{4} \quad (7.14)$$

$$D_i = b_i D_{i-1} - \frac{b_i^2}{4} D_{i-2} \quad 3 \leq i \leq n.$$

$\mathbf{D}_i > 0$ if $\forall 1 \leq j \leq i$ $D_j > 0$. This implies the following inequality relations for b_i's:

$$b_1 > 0$$

$$b_2 < 4b_1 \quad (7.15)$$

$$b_i < \frac{4D_{i-1}}{D_{i-2}}$$

$$3 \leq i \leq n.$$

Thus if we choose b_i such that the inequalities (7.15) are satisfied then $\mathbf{D} > 0$ (positive definate) and the system is stable. This in-turn implies that the section lengths L_i have to be chosen such that the inequalities (7.15) are satisfied.

Stability in the Equal Section Case

In this section we will study the stability of the linear system given by (7.11) under the equal section case. Under the equal section case $b_i = \frac{L}{(L/n)} = n$. Thus the reccurence relation (7.14) assumes the following form:

$$D_1 = b_1 = n$$

$$D_2 = b_1 b_2 - \frac{b_2^2}{4} = \frac{3n^2}{4} \tag{7.16}$$

$$D_i = b_i D_{i-1} - \frac{b_i^2}{4} D_{i-2} = n(D_{i-1} - \frac{n}{4} D_{i-2}) \quad 3 \leq i \leq n.$$

By simple mathematical induction we can prove that:

$$D_i = n^i \frac{i+1}{2^i} > 0 \quad 1 \leq i \leq n \tag{7.17}$$

This means that for $b_i = n$ $\forall$ $1 \leq i \leq n$ the system is asymptotically stable.

7.3.2 Saturation of Control

It can be seen from (7.3) that the flow velocity v is directly proportional to the free flow velocity which is the control parameter in the state equations (7.7). In fact its the velocity of a pedestrian under the situation where the density $\rho = 0$. Since there is a upper limit on how fast even a single pedestrian can move hence there is a upper limit on the parameter v_f. This translates to an upper limit on the control parameters v_{f_i} in the state equations (7.7). Since the control parameters are already normalized with respect to the length of the corridor hence the saturation limit $\bar{v}_f$ represents the maximum number of lengths of the corridor that can be covered in a unit time. From (7.9) this translates mathematically to the following inequality:

$$v_{f_i} = \frac{k}{1 - \rho_i} \leq \bar{v}_f \tag{7.18}$$

Readjusting the inequalities (7.18) results in the following inequalities in terms of the state variables to ensure non-saturation of control:

$$\rho_i \leq \left(1 - \frac{k}{\bar{v}_f}\right) = \mu \quad 1 \leq i \leq n. \tag{7.19}$$

7.4 Simulation Results

Conditions (7.19) are sufficient for keeping all the control variables v_{f_i} $1 \leq i \leq n$ under the saturation limit. To ensure that the state constraints (7.8) are satisfied choose k in (7.19) such that:

$$k = \alpha \bar{v}_f \quad 0 \leq \alpha \leq 1 \tag{7.20}$$

This results in the following value for μ in (7.19):

$$\mu = (1 - \alpha). \tag{7.21}$$

Theorem 7.3.1 *If $\rho_i(0) \leq \mu$ then $\rho_i(t) \leq \mu$ for $0 \leq t \leq \infty$. i.e. if the control is initially unsaturated then it is always unsaturated.*

Proof

$$\dot{\rho}_1 = -kb_1\rho_1$$
$$\dot{\rho}_i = kb_i\rho_{i-1} - kb_i\rho_i \quad 2 \leq i \leq n \tag{7.22}$$

for $k = 1$

$$\dot{\rho}_1 = -kb_1\rho_1 \leq 0$$

This implies

$$\rho_1(t) \leq \rho_1(0) \leq \mu \quad \forall \quad 0 \leq t \leq \infty \tag{7.23}$$

Let $\rho_k = \mu$ for some $2 \leq k \leq n$ & $\rho_j \leq \rho_k$ $2 \leq j \leq n$ & $k \neq j$ at $t = \tau$ this implies

$$\dot{\rho}_k = kb_k(\rho_{k-1} - \rho_k) = -a \leq 0 \tag{7.24}$$

for $a \geq 0$ this implies

$$\rho_k(\tau + \delta\tau) = \rho_k(\tau) - a\delta\tau \leq \mu \tag{7.25}$$

for $\delta\tau \to 0$

Hence $\forall$ $2 \leq k \leq n$ $\rho_k(t) \leq \mu$ $\forall$ $0 \leq t \leq \infty$

7.4 Simulation Results

A comparison is made between what would be an obvious choice in a panic situation for the evacuees under an emergency evacuation order and the feedback control developed. It is also shown later that if the feedback control developed is discretized still the decay profile of the densities is not affected much. The obvious choice made by the pedestrians without guidance would be to move at the fastest possible free flow speed. In terms of the system (7.7) developed this would mean that the control parameters v_{f_i} are operating at a saturation value. A sample case is chosen such that the normalized pedestrian densities in all the sections are equal and are very close to μ. The densities chosen are:

$$\rho_i = 0.9\mu \quad 1 \leq i \leq n \qquad (7.26)$$

A corridor of length 50 m is assumed and divided into three sections. A non-normalized saturation free flow speed is assumed to be $v_{f_s} = 4m/s$. This would correspond to a normalized saturation free flow speed of $\bar{v}_f = 0.08\,\mathrm{s}^{-1}$. The α value in (7.20) is chosen to be sufficiently small so as to get a reasonably high value for μ. $\alpha = 0.1$ is chosen so that $\mu = 0.9$. Figure 7.2 shows the evolution of the normalized pedestrian densities if the evacuees were to move at a saturation free flow speed of $v_{f_s} = 4m/s$ having a normalized value of $\bar{v}_f = 0.08\,\mathrm{s}^{-1}$. It is clearly seen from the plot that within 8 s of the start of the evacuation the normalized pedestrian density in the

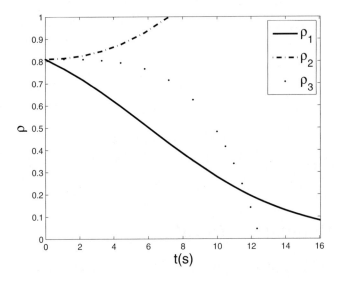

Fig. 7.2. Evolution of the normalized pedestrian densities under saturation control

7.4 Simulation Results

2nd section reaches a critical value of 1 which implies from (7.3) that the flow velocity, $v = 0$ and jamming occurs. This is highly undesirable as not only is the evacuation process stopped but there is an increased possibility of panic and could result in something as catastrophic as a stampede. Hence there is a high motivation for controlling the free flow velocity such that a smooth transition of the pedestrian densities in all the sections from their initial values to empty state of the corridor is ensured. This controlling mechanism also has to ensure that the free flow velocity remains within the saturation value. Figure 7.3 shows the evolution of the normalized pedestrian densities under the proposed feedback control for the same initial conditions.

It can be clearly seen that the pedestrian densities asymptotically approach 0 and there is no possibility of reaching jam density anywhere. Figure 7.4 show the evolution of the normalized control free flow velocities.

From these plots it is evident that the normalized control free flow velocities are well below their saturation value at all times in all the sections. Hence the objective of non-saturation is easily obtained without having to artificially impose the saturation constraint.

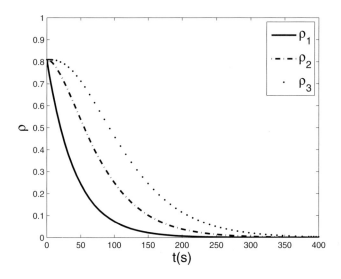

Fig. 7.3. Evolution of the normalized pedestrian densities under feedback linearized control

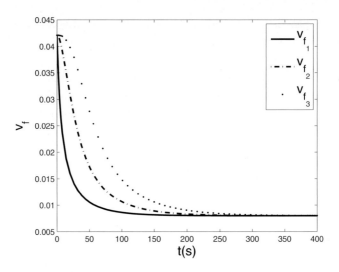

Fig. 7.4. Control free flow velocity for continuous feedback control

In reality however the continuous form of the control equation (7.9) has to be implemented in a discretized form. Hence a simulation was run to see how the performance would be under the discretized implementation of this control. It was assumed for implementation

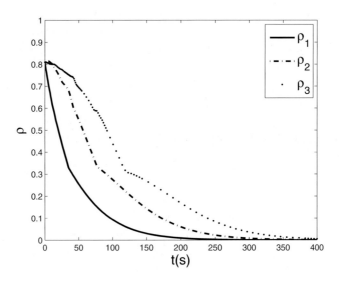

Fig. 7.5. Evolution of the normalized pedestrian densities under discretized feedback control

7.5 Code

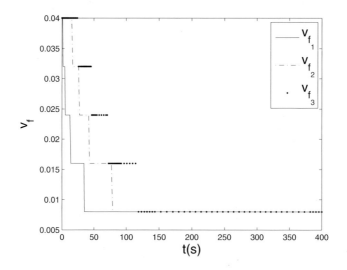

Fig. 7.6. Discretized normalized control free flow velocities in sections 1,2 and 3

purposes that the control in the from of normalized free flow velocity can take 10 distinct values between 0 and $\bar{v}_f$. The feedback control was computed using (7.9) and was rounded off to the nearest control value that could be implemented. Figure 7.5 shows the evolution of the normalized pedestrian densities in all the three sections under discretized feedback control implementation. Figure 7.6 shows the control free flow velocities in all the three sections.

A comparison of Fig. 7.3 and 7.5 shows similar evolution of state profiles. Thus the control proposed in (7.9) has a good feasibility of practical implementation. In the next section the matlab code written to perform the above simulations is presented.

7.5 Code

In this section the Matlab code used for the above simulation is described. This code can be found in the section Computer Code 7.7 after the exercises. The idea is to explain the numerical implementation with the help of a working Matlab code. The code is divided into three files:

1. main.m
2. rhodot_nsec.m
3. vfcntrl_nsec_try.m

The file main.m sets up the parameters for the simulation. It sets up the number of sections, the interval of simulation, assigns values to α, b_i and $\bar{v}_f$ and assigns initial values to the pedestrian densities in each section. It then uses various control options in the variable cntrl to perform simulations and obtain results for the states ρ in terms of the time t. To do this it uses the well established *ode* solver *ODE45* which uses the routine rhodot_nsec to compute the derivatives of the state. The routine rhodot_nsec computes the derivatives of the states using the state equations in (7.7). To compute the derivative it needs the control values v_f which it acquires from the routine vfcntrl_nsec_try. The routine vfcntrl_nsec_try has a variable cntrl which decides on the type of control to be used. cntrl=1 corresponds to saturation control, cntrl=2 corresponds to feedback linearizing control and cntrl=3 corresponds to discretized feedback linearizing control.

7.6 Exercises

Problem 7.6.1 Instead of using the *Greenshields* model as the $v-\rho$ model use the following models to derive the feedback linearizing control in (7.9):

1. *Pipes-Munjal Model*: $v = v_f \left(1 - \left(\frac{\rho}{\rho_m}\right)^n\right)$

2. *Greenberg Model*: $v = v_f \ln\left(\frac{\rho}{\rho_m}\right)$

3. *Underwood Model*: $v = v_f \exp\left(-\frac{\rho}{\rho_m}\right)$

Problem 7.6.2 In the simulation for the length of the corridor $L = 50\ m$ use unequal section lengths L_i, $i = 1, 2, 3$ and see if there is any performance improvement over the equal section case?

Problem 7.6.3 Repeat the above simulation for four sections and see if there is any performance improvement?

7.7 Computer Code

Problem 7.6.4 Adjust the value of α such that the gain increases while the initial densities ρ_{i0} can be kept the same as above. Find the performance improvement in this case?

7.7 Computer Code

7.7.1 main.m

```
% Code to run the simulations for various
% controls and plot the results for
% pedestrian evacuation

% Setting up the parameters

alpha=0.1 ;
n=3 ; % n=number of sections of corridor;
tspan=[0,400]; % interval of simulation
vfm=0.08; % Normalized saturation free
          % flow velocity
b=ones(n,1)*n; % b=vector of b_i.
k=alpha*vfm ; % k=feedback linearization
              % gain
mu=1-alpha ; % mu=upper limit on the states
             % to avoid saturation of control

% rhoeq is the equal rho assumed in
% each section
rhoeq=0.9*mu ;
rho0=ones(n,1)*rhoeq ;

cntrl=1 ; % simulation for saturation
          % control
[t,rho]=ode45(@rhodot_nsec,tspan,rho0,[],vfm...
,b,k,cntrl) ;

cntrl=2 ; % simulation for feedback
          % linearizing control
[tcnt,rhocnt]=ode45(@rhodot_nsec,...
```

```
tspan,rho0,[],vfm,b,k,cntrl) ;

cntrl=3 ; % simulation for discretized
          % feedback linearizing control

[tdis,rhodis]=ode45(@rhodot_nsec,...
tspan,rho0,[],vfm,b,k,cntrl) ;

% recomputing the control from the
% state values.

% Continuous Control
for it=1:1:length(tcnt)
rhot=rhocnt(it,:) ;
rhot=rhot' ; % col. vect. of states at t
cntrl=2 ; % feedback linearizing control
vft=vfcntrl_nsec_try(rhot,vfm,k,cntrl) ;
% vft= free flow velocity control for
% feedback linearizing control
vfcnt(it,:)=vft' ;
end

% Discretized Control
for it=1:1:length(tdis)
rhot=rhodis(it,:) ;
rhot=rhot' ; % col. vect. of states at t
% discretized feedback linearizing control
cntrl=3 ;
% vft= free flow velocity control for
% disc. feedback linearizing control
vfdis(it,:)=vft' ;
end

% Plotting of Results

% rho v/s t for saturation control
figure ;
rsp=plot(t,rho(:,1),'k-',t,rho(:,2),...
```

7.7 Computer Code

```
       'k-.',t,rho(:,3),'k.') ;
set(rsp,'LineWidth',2) ;
h=legend('\rho_1','\rho_2','\rho_3') ;
set(h,'FontSize',16) ;
xlabel('t(s)','FontSize',16) ;
ylabel('\rho','FontSize',16) ;
axis([0 t(end) 0 1]) ;

% rho v/s t for feedback lin. control
figure ;
rcp=plot(tcnt,rhocnt(:,1),'k-',tcnt,...
    rhocnt(:,2),'k-.',tcnt,rhocnt(:,3),'k.');
set(rcp,'LineWidth',2) ;
h=legend('\rho_1','\rho_2','\rho_3');
set(h,'FontSize',16) ;
xlabel('t(s)','FontSize',16) ;
ylabel('\rho','FontSize',16) ;
axis([0 tcnt(end) 0 1]) ;

% vf v/s t for feedback lin. control
figure ;
vfp=plot(tcnt,vfcnt(:,1),'k-',tcnt,...
    vfcnt(:,2),'k-.',tcnt,vfcnt(:,3),'k.') ;
set(vfp,'LineWidth',2) ;
h=legend('v_{f_1}','v_{f_2}','v_{f_3}');
set(h,'FontSize',16);
xlabel('t(s)','FontSize',16);
ylabel('v_f','FontSize',16) ;

% rho v/s t for disc. feedback lin. control
figure ;
rcdp=plot(tdis,rhodis(:,1),'k-',tdis,...
    rhodis(:,2),'k-.',tdis,rhodis(:,3),'k.') ;
set(rcdp,'LineWidth',2) ;
h=legend('\rho_1','\rho_2','\rho_3');
set(h,'FontSize',16) ;
xlabel('t(s)','FontSize',16);
ylabel('\rho','FontSize',16) ;
```

```
axis([0 tdis(end) 0 1]) ;

% vf v/s t for disc. feedback lin. control
figure ;
vfdp=plot(tdis,vfdis(:,1),'k-',tdis,...
    vfdis(:,2),'k-.',tdis,vfdis(:,3),'k.') ;
xlabel('time') ; ylabel('v_f_o') ;
legend('v_f_1','v_f_2','v_f_3') ;
set(vfp,'LineWidth',2) ;
h=legend('v_{f_1}','v_{f_2}','v_{f_3}');
set(h,'FontSize',16) ;
xlabel('t(s)','FontSize',16);
ylabel('v_f','FontSize',16) ;

return ;
```

7.7.2 rhodot_nsec.m

```
% Routine: rhodot_nsec

function rhod=rhodot_nsec(t,rho,vfm,b,k,cntrl)
n=length(rho) ; % n sections
% Routine to compute the control
vf=vfcntrl_nsec_try(rho,vfm,k,cntrl) ;
% time derivative of rho(1)
rhod(1,1)=-rho(1)*(1-rho(1))*b(1)*vf(1,1) ;
% column vector of time
% derivative of rho(2:n)
rhod(2:n,1)=rho(1:n-1).*(1-rho(1:n-1))...
.*b(2:n,1).*vf(1:n-1,1)-rho(2:n)...
.*(1-rho(2:n)).*b(2:n,1).*vf(2:n) ;
```

7.7.3 vfcntrl_nsec_try.m

```
function vf=vfcntrl_nsec_try(rho,vfm,k,cntrl)
switch cntrl
    case 1
% saturation control
    n=length(rho) ; % number of sections
```

7.7 Computer Code

```
    vf(:,1)=vfm*ones(n,1) ; % saturation
    % value in each section
    case 2
%feedback linearizing control
    vf(:,1)=k./(1-rho) ;
% Feedback linearized value in each section
    case 3
% discrete feedback linearizing control
    vf(:,1)=k./(1-rho) ;
% Feedback linearized value computed in each
% section
    vfs=vfm/10 ; % interval of discretization
    nvf=round(vf/vfs) ;
%nvf=vector of integers created by rounding
% off the
% ratio vf_i/vfs for each section i
    vf=nvf*vfs ;
% vf=vector of discretized feedback
% linearized control
% values
end
```

Chapter 8
Discretized Optimal Control

Evacuation Dynamics and Control is a field of critical importance given the threats of modern urban life. Evacuation can be classified as broadly belonging to two types: (1) Evacuation from a built up facility and (2) Evacuation from a locality. While the first kind involves moving pedestrians out of the buildings quickly the second type mainly involves appropriate routing of vehicular traffic. This paper involves optimal evacuation of pedestrians from the building facilities. Pedestrian dynamics has been studied using two types of models (1) macroscopic and (2) microscopic. The macroscopic models typically deal with the pedestrian flow as a continuum where as the microscopic models deal with the movement and interactions of the individual pedestrians. Various cellular automata and simulation based models are available to deal with the pedestrian flow at the microscopic level [42, 47, 49, 50]. At the macroscopic level the pedestrian flow is analyzed in terms of the variables like pedestrian flow density and flow velocity. Various models developed for macroscopic traffic flow (Chap. 2 of [57]) can be readily extended to capture the dynamics of pedestrian flow. The dynamics of pedestrian flow described at the macroscopic level can be utilized to develop control for quick evacuation. Feedback controls for evacuation have been developed based on macroscopic models in [99] for the 1-D case and in [100] for the 2-D case. A wealth of literature exists dealing with dynamic traffic assignment [57, 81, 96]. A Building structure can be considered to be a network of rooms, corridors, stairways and exits. While the rooms and exits can be thought of as the sources and sinks of the pedestrian flow respectively, the corridors and stairways can be thought of as circulation elements which will be the key elements in routing the pedestrians out of the buildings. A controlled flow of pedestrians in these links can achieve a fast evacuation of the pedestrians from these links and hence from the entire facility. Studies have also been done to design network links of proper dimensions to accommodate a quick evacuation using queuing theory [5].

8.1 Optimal Control

In this section we state the necessary conditions for the optimal control of pedestrian flow described by (7.6) derived for the pedestrian flow model in Chap. 7. In Sect. 8.1.1 we give the state equations and there by define the control variables. In Sect. 8.1.2 we define the cost function that we intend to minimize. Then using the state equations and the cost function we develop the necessary conditions for the optimal control in Sect. 8.1.3 by the calculus of variations method described in Chap. 5 of [63]. The problem considered here is that of free final state and fixed final time.

8.1.1 State Equations

For the purpose of developing an optimal control for the system of (7.6) in Chap. 7 we write them in the following form by dropping the tildes. Henceforth the normalized quantities will be represented without the tildes.

$$\dot{\rho}_1 = -\rho_1(1-\rho_1)b_1 v_{f_1} \tag{8.1}$$

$$\dot{\rho}_i = \rho_{i-1}(1-\rho_{i-1})b_i v_{f_{i-1}} - \rho_i(1-\rho_i)b_i v_{f_i} \tag{8.2}$$

$$\dot{v}_{f_i} = u_i \tag{8.3}$$

The control variable u is introduced in the last (8.3). As is apparent from the equation the control input represents the time rate

8.1 Optimal Control

of change of free flow velocities in the various sections. State vector in this formulation is

$$\mathbf{x} = \begin{bmatrix} \rho \\ \mathbf{v_f} \end{bmatrix} \tag{8.4}$$

We collect the right hand sides of (8.1), (8.2) and (8.3) in the vector $\mathbf{a}(\mathbf{x}(t), \mathbf{u}(t), t)$ given as follows.

$$a_1(\mathbf{x}(t), \mathbf{u}(t), t) = -\rho_1(1-\rho_1)b_1 v_{f_1} \tag{8.5}$$

$$a_i(\mathbf{x}(t), \mathbf{u}(t), t) = \rho_{i-1}(1-\rho_{i-1})b_i v_{f_{i-1}} - \rho_i(1-\rho_i)b_i v_{f_i} \tag{8.6}$$
$$2 \le i \le n$$

$$a_{v_{f_i}}(\mathbf{x}(t), \mathbf{u}(t), t) = u_i \tag{8.7}$$
$$1 \le i \le n$$

This results in the following state equations:

$$\dot{\mathbf{x}} = \mathbf{a}(\mathbf{x}(t), \mathbf{u}(t), t). \tag{8.8}$$

8.1.2 Cost Function

For the development of the optimal control we use the following as the cost function:

$$J(\mathbf{u}(t)) = \int_{t_0}^{t_f} g(\mathbf{x}(t), \mathbf{u}(t), t) d\tau + h(\mathbf{x}(t_f), t_f), \tag{8.9}$$

where

$$g(\mathbf{x}(t), \mathbf{u}(t), t) = \sum_{i=1}^{n} \rho_i^2 + u_i^2 \tag{8.10}$$

and

$$h(\mathbf{x}(t_f), t_f) = \sum_{i=1}^{n} \rho_i(t_f)^2. \tag{8.11}$$

Here, $h(\mathbf{x}(t_f), t_f)$ represents the terminal cost. By defining the cost function by (8.9) we are assured that its minimization will ensure low pedestrian density as well as low rate of change of free flow velocity at every instant of time. Since both the attributes are highly desirable from the point of view of evacuation, the particular choice of cost function given by (8.9) is justified.

8.1.3 Calculus of Variation

We now use the calculus of variations approach to develop the Euler Lagrange equations that need to be satisfied by the optimal control and the corresponding states. The Hamiltonian for the equations

$$\dot{\mathbf{x}} = \mathbf{a}(\mathbf{x}, \mathbf{u}, t)$$

is given by

$$H(\mathbf{x}(t), \mathbf{u}(t), \mathbf{p}(t), t) = g(\mathbf{x}(t), \mathbf{u}(t), t) + \mathbf{p}^T \mathbf{a}(\mathbf{x}(t), \mathbf{u}(t), t). \quad (8.12)$$

In the above equation the vector $\mathbf{p}$ is the *co-state* vector. For (8.1), (8.2) and (8.3) the Hamiltonian is given by

$$\begin{aligned} H(\mathbf{x}(t), \mathbf{u}(t), \mathbf{p}(t), t) = & \sum_{i=1}^{n}(\rho_i^2 + u_i^2) \\ & + p_1(-\rho_1(1-\rho_1)b_1 v_{f_1}) \\ & + \sum_{i=2}^{n} p_i(\rho_{i-1}(1-\rho_{i-1})b_i v_{f_{i-1}}) \\ & - \rho_i(1-\rho_i)b_i v_{f_i}) + \sum_{i=1}^{n} p_{v_{f_i}} u_i \end{aligned} \quad (8.13)$$

Necessary Conditions

We now state Euler Lagrange equations which specify the necessary conditions for the control to be a optimal

$$\dot{\mathbf{x}}^* = \frac{\partial H}{\partial \mathbf{p}}(\mathbf{x}^*, \mathbf{u}^*, \mathbf{p}^*, t) \quad (8.14)$$

These (8.14) are the state equations given by (8.1), (8.2) and (8.3). These equations are stated collectively in terms of the vector

$$x = \begin{bmatrix} \rho \\ \mathbf{v_f} \end{bmatrix},$$

in (8.8).

8.1 Optimal Control

$$\dot{\mathbf{p}}^* = -\frac{\partial H}{\partial x}(\mathbf{x}^*, \mathbf{u}^*, \mathbf{p}^*, t). \tag{8.15}$$

Equation (8.15) gives the co-state equations that are specified below:

$$\dot{p}_i = \begin{pmatrix} -(2\rho_i + p_i(-1 + 2\rho_i)b_i v_{f_i} + \\ p_{i+1}(1 - 2\rho_i)b_{i+1} v_{f_i}) \end{pmatrix}, \tag{8.16}$$

$$1 \leq i \leq n-1,$$

$$\dot{p}_n = -(2\rho_n + p_n(-1 + 2\rho_n)b_n v_{f_n}), \tag{8.17}$$

$$\dot{p}_{v_{f_i}} = -(p_i(-\rho_i(1-\rho_i)b_i) + p_{i+1}(\rho_i(1-\rho_i)b_{i+1})), \tag{8.18}$$

$$1 \leq i \leq n-1,$$

$$\dot{p}_{v_{f_n}} = -(p_n(-\rho_n(1-\rho_n)b_n)). \tag{8.19}$$

Another necessary condition on the Hamiltonian is that its partial derivative with respect to the control must vanish for all times $t_0 \leq t \leq t_f$.

$$\frac{\partial H}{\partial \mathbf{u}}(\mathbf{x}^*, \mathbf{u}^*, \mathbf{p}^*, t) = 2u_i^* + p_{v_{f_i}}^* = 0 \tag{8.20}$$

We choose the case in which final time t_f is specified and the final state $\mathbf{x}(t_f)$ is free. Hence we get the following boundary conditions. At the initial time t_0 we have,

$$\rho^*(t_0) = \rho_0 \tag{8.21}$$

$$\mathbf{v}_{\mathbf{f}_i}^*(t_0) = 0 \tag{8.22}$$

At the final time t_f we have,

$$\mathbf{p}^*(t_f) = \frac{\partial h}{\partial \mathbf{x}}(\mathbf{x}^*(t_f), t_f) \tag{8.23}$$

$$p_i^*(t_f) = 2\rho_i^*(t_f) \tag{8.24}$$

$$p^*_{v_{f_i}} = 0 \tag{8.25}$$

This is a typical two point boundary value problem that occurs in optimal control. The boundary conditions are a part of the necessary conditions.

8.2 The Method of Steepest Descent

The Method of Steepest Descent was used to compute the optimal control in a piecewise constant manner. This is described in Chap. 6 of [63]. This is an iterative method in which the control profile is updated in every iteration at every discretized time instant at which it was initially assumed. An initial piecewise constant control profile is assumed and the states are computed by forward integration. The the boundary conditions for the co-state $p^*_i(t_f)$ are computed using (8.24) and (8.25). Then (8.16), (8.17), (8.18) and (8.19) are integrated backwards in time from $t = t_f$ to $t = t_0$. Using the co-state values $\partial H(t)/\partial \mathbf{u}$ is calculated using the first equality in (8.20) at every discrete time instant at which the control value $\mathbf{u}$ is assumed. Since the control profile was arbitrarily chosen this in general will not be zero. As the necessary condition (8.20) requires this quantity to be zero, the control will have to be updated in the direction of the steepest descent of the Hamiltonian H at every discrete time instant. This corresponds to the negative gradient of H with respect to $\mathbf{u}$, $-\partial H/\partial \mathbf{u}$. Hence we have the following control update law at every iteration

$$\mathbf{u}^{(i+1)} = \mathbf{u}^{(i)} - \tau \frac{\partial H^{(i)}}{\partial \mathbf{u}} \tag{8.26}$$

τ in (8.26) has to be chosen such that the cost function in (8.9) has to continually decrease in every iteration. This usually is done by an ad hoc strategy. The iterations stop when

$$\left\| \frac{\partial H^{(i)}}{\partial \mathbf{u}} \right\| \leq tol_1 \tag{8.27}$$

or

$$|(J^{(i+1)} - J^i)| \leq tol_2, \tag{8.28}$$

where tol_1 and tol_2 are pre-specified tolerances.

8.3 Numerical Results

A test case was considered for the above optimal control formulation. A corridor was divided into two sections and each was assigned a initial normalized pedestrian density of $\rho_{i0} = 0.6$. The initial free flow velocities were set to $v_{f_{i0}} = 0$ as the crowd is stationary initially. The sections are taken to be of equal length, so $b_i = 2$. The initial time and final times are $t_0 = 0$ and $t_f = 10\,\mathrm{s}$ respectively. The interval $[t_0, t_f]$ was divided into subintervals of $0.1\,\mathrm{s}$. The method of steepest descent described above was used to obtain the optimal control. A matlab code was written to solve the two-point boundary value problem above and the plots were obtained for ρ_i, vf_i, u_i, p_i, p_{vf_i} versus time. A simple euler solver was used to solve the ode's involved.

It can be seen in Fig. 8.2 that almost all of the population is evacuated in a period of 10 s. From Fig. 8.6 it is apparent that the value of $\partial H / \partial u_i$ is very close to zero. This was forced using a low value of tolerance, tol_1 in (8.27). Thus the requirement of optimality in (8.20) is closely satisfied. In reality when implementing this control strategy we can give instructions based on the profile of $\mathbf{v_f}$ in Fig. 8.3 rather than $\mathbf{u}$ in Fig. 8.1.

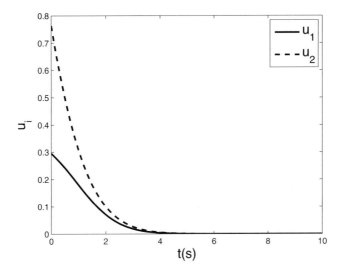

Fig. 8.1. Optimal control u_1 and u_2 as a function of time, t

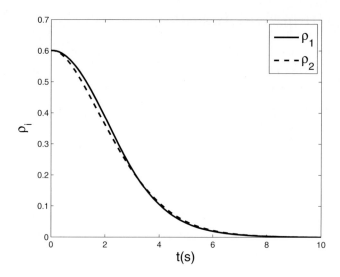

Fig. 8.2. Evolution of pedestrian densities ρ_1 and ρ_2 as a function of time, t

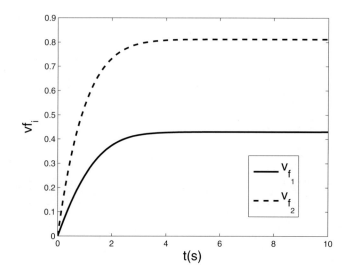

Fig. 8.3. Evolution of free flow velocities v_{f_1} and v_{f_2} as a function of time, t

8.3 Numerical Results

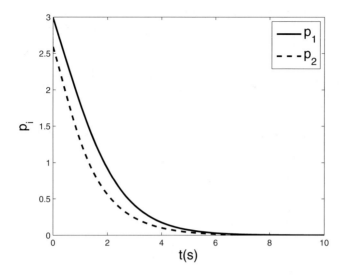

Fig. 8.4. Evolution of co-state, p_1 and p_2 as a function of time, t

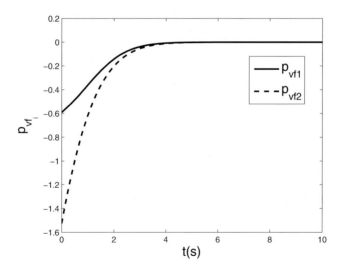

Fig. 8.5. Evolution of co-state, $p_{v_{f_1}}$ and $p_{v_{f_2}}$ as a function of time, t

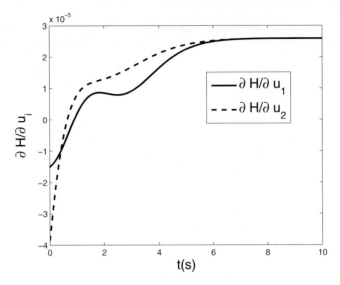

Fig. 8.6. $\frac{\partial H}{\partial u_i}$ as a function of time, t

8.4 Code

In this section the code written to perform the above simulation is explained. It is written in matlab and consists of two m-files. The m-files are:

1. main.m

2. optimal_cntrl_calc_var_nsec_odesol.m

The idea of this section is to explain the numerical *method of steepest descent* described above with a working code written in matlab. In the following sections the workings of the above mentioned m-files are described.

8.4.1 main.m

This file performs the task of setting up the parameters namely `tvec` which is a column vector consisting of time instants at which the control profile is assumed and sought, `st0` the initial states and `b` which is a vector of L/L_i. After this the routine `optimal_cntrl_calc_var _nsec_odesol` is called for computing the optimal control, `utim`,

8.4 Code

the corresponding states stmat, co-states, costmat, $\partial H/\partial u$, dhdu, a vector consisting of the cost J in each iteration Jiter and the Hamiltonian corresponding to the optimal control H at each time instant in tvec.

8.4.2 optimal_cntrl_calc_var_nsec_odesol.m

This file contains a function with the same name which uses the *method of steepest descent* to compute the optimal control at each instant of time in tvec. It uses the following functions for this purpose:

1. [stmat,utim,J,test] = cntrl_update (tvec,stmat,utim, dhdu,b,gam1,gam2): This routine performs the contorl update and computes the corresponding states and co-states.

2. stmat = fwdstinteg(tvec,utim,st0,b): This routine performs the forward integration to compute the states.

3. stdot = stdyn(st,b,u): This routine computes the time derivative of the state vector.

4. costmat = bkcostinteg(tvec,stmat,costf,b): This routine performs the backward integration to compute the co-states

5. costdot = costdyn(cost,b,st): This routine computes the time derivative of the co-state vector.

An initial control profile is assumed in the variable utim and the states stmat, are computed by forward integration using the routine fwdstinteg. Using this value of utim and stmat the cost function J is computed. After this the tolerences tol1 and tol2 are specified which determine when the iterations updating the control are to stop. The iterations updating the control stop when $\|\partial H^{(i)}/\partial u\| \leq tol_1$ or $|(J^{(i+1)} - J^i)| \leq tol_2$. The variable test determines whether the above two requirements are met. test==0 indicates that the convergence is not achieved while test==1 indicates that convergence has been achieved. In the iteration loop the co-states are initially computed by backward integration of the co-state equations. This is done using the routine bkcostinteg. Next the gradient of the Hamiltonian wrt the control $\partial H/\partial u$ is computed in the matrix dhdu. Subsequently the control update step is done by the routine cntrl_update. This routine outputs the updated control utim and the states stmat and the cost J corresponding to these. The routine cntrl_update also assigns a value to the variable test after checking the convergence from the criteria described above. Thus after checking the value of the variable test it is decided whether to proceed with the next iteration or not. In the next section the interface to the code routines is explained.

8.5 Exercises

Problem 8.5.1 Instead of using the *Greenshields* model as the $v - \rho$ model use the following models to derive the state equations (8.1), (8.2),(8.3), the co-state equations (8.16), (8.17), (8.18) and (8.19) and the necessary condition (8.20)

1. Pipes-Munjal Model: $v = v_f \left(1 - \left(\frac{\rho}{\rho_m}\right)^n\right)$
2. Greenberg Model: $v = v_f \ln\left(\frac{\rho}{\rho_m}\right)$
3. Underwood Model: $v = v_f \exp\left(-\frac{\rho}{\rho_m}\right)$

Problem 8.5.2 Obtain the control, state and the co-state profiles for the above models for the 2 section case using the method of steepest descent described in Sect. 8.2. In case you intend to use the code given you only need to modify the routines *stdyn* and *costdyn*.

Problem 8.5.3 Try to see if changing the b values can improve the optimal performance time. If you are using the code given you only need to change the value of b at multiple places.

8.6 Computer Code

8.6.1 optimal_corridor_evacuation/main

```
% Setting up the parameters
%time instants for simulation
tvec=[0:0.1:10]';
st0=[1;1;0;0]*0.6; % initial state
b=2*ones(2,1); % b values
```

8.6 Computer Code

```
% Calling the optimal control routine
[stmat,utim,costmat,dhdu,Jiter,H]...
=optimal_cntrl_calc_var_nsec_odesol...
(tvec,st0,b) ;

% Plotting the results
% rho=ped. densities, vf= free flow
% velocities
rho=stmat(:,1:2) ; vf=stmat(:,3:4) ;
% plot of (u_1,u_2) v/s time
figure ;
hp=plot(tvec,utim(:,1),'k-',...
    tvec,utim(:,2),'k--');
set(hp,'LineWidth',2) ;
h=legend('u_1','u_2');
set(h,'FontSize',16) ;
xlabel('t(s)','FontSize',16) ;
ylabel('u_i','FontSize',16) ;

% plot of (rho_1,rho_2) v/s time
figure ;
hp=plot(tvec,stmat(:,1),'k-',...
    tvec,stmat(:,2),'k--');
set(hp,'LineWidth',2) ;
h=legend('\rho_1','\rho_2');
set(h,'FontSize',16) ;
xlabel('t(s)','FontSize',16) ;
ylabel('\rho_i','FontSize',16) ;

% plot of (vf_1,vf_2) v/s time
figure ;
hp=plot(tvec,stmat(:,3),'k-',...
tvec,stmat(:,4),'k--');
set(hp,'LineWidth',2) ;
h=legend('v_{f_1}','v_{f_2}');
set(h,'FontSize',16) ;
xlabel('t(s)','FontSize',16) ;
ylabel('vf_i','FontSize',16) ;
```

```
% plot of (p_1,p_2) v/s time
figure;
hp=plot(tvec,costmat(:,1),...
'k-',tvec,costmat(:,2),'k--');
set(hp,'LineWidth',2);
h=legend('p_1','p_2');
set(h,'FontSize',16) ;
xlabel('t(s)','FontSize',16) ;
ylabel('p_i','FontSize',16) ;

% plot of (p_vf1,p_vf2) v/s time
figure ;
hp=plot(tvec,costmat(:,3),'k-'...
,tvec,costmat(:,4),'k--');
set(hp,'LineWidth',2);
h=legend('p_{vf1}','p_{vf2}');
set(h,'FontSize',16) ;
xlabel('t(s)','FontSize',16) ;
ylabel('p_{vf_i}','FontSize',16)

% plot of (dh/du_1,dh/du_2) v/s
% time
figure ;
hp=plot(tvec,dhdu(:,1),'k-',...
tvec,dhdu(:,2),'k--') ;
set(hp,'LineWidth',2);
h=legend('{\partial H}/{\partial u_1}'...
    ,'{\partial H}/{\partial u_2}') ;
set(h,'FontSize',16) ;
xlabel('t(s)','FontSize',16) ;
ylabel('{\partial H}/{\partial u_i}'...
    ,'FontSize',16) ;

% plot of hamiltonian v/s time
figure ;
plot(tvec,H) ; legend('H') ;
```

8.6 Computer Code

8.6.2 optimal_cntrl_calc_var_nsec_odesol

```
function [stmat,utim,costmat,dhdu,Jiter,H]...
=optimal_cntrl_calc_var_nsec_odesol(tvec,st0,b)
% This function computes the optimal control and
% the corresponding states and the co-states.
% Method of Steepest Descent is used for this.

% n=no. of sections
nt=length(tvec) ; n=length(st0) ; n=n/2 ;
utim=0.1*ones(nt,n) ; % assume control
% forward integration for states
stmat=fwdstinteg(tvec,utim,st0,b) ;
rho=stmat(:,1:n); rhof=rho(end,:);
% time increment for simulation
dt=tvec(2)-tvec(1) ;
% cost function for the simulation
J=sum(sum(rho.^2+utim.^2,2)*dt)+sum(rhof.^2);
% iter=no of control update iterations
% Jiter=vector of iteration cost values
iter=1 ; Jiter(iter)=J ;
% convergence:
% test=0 => not occured  test=1 => occured.
test=0 ;
% tolerences
tol1=10^-4 ; tol2=10^-6 ;
% iterations for control and cost till
% convergence occurs
while test==0 % check for convergence
iter=iter+1 ;
stf=stmat(end,:)' ; rhof=stf(1:n) ;
costf=[2*rhof;zeros(n,1)] ;
% backward integration for co-states
costmat=bkcostinteg(tvec,stmat,costf,b) ;
% calculation dhdu at all time
dhdu=2*utim+costmat(:,n+1:2*n) ;
% "cntrl_update" updates the control and state
% for the next iteration
[stmat,utim,J,test]=...
```

```
   cntrl_update(tvec,stmat,utim,dhdu,b,tol1,tol2) ;
% Jiter(iter)=value of the cost function at the
% current iteration.
Jiter(iter)=J ;
end
rho=stmat(:,1:n) ; vf=stmat(:,n+1:2*n) ;
p=costmat(:,1:n) ; pvf=costmat(:,n+1:2*n) ;
% Computation for hamiltonian
H=zeros(nt,1) ;
for i=1:1:n
H=H+rho(:,i).^2+utim(:,i).^2+p(:,i)...
.*(-rho(:,i).*(1-rho(:,i))*b(i).*vf(:,i)) ;
H=H+pvf(:,i).*utim(:,i) ;
if i>=2
H=H+p(:,i).*(rho(:,i-1).*(1-rho(:,i-1))...
    *b(i).*vf(:,i-1)) ;
end
end
end

function [stmat,utim,J,test]=...
cntrl_update(tvec,stmat,utim,dhdu,b,tol1,tol2)
test=0 ;
dt=tvec(2)-tvec(1); nsi=size(stmat);
% n=no of sections, st0=initial states
n=nsi(2); n=n/2; st0=stmat(1,:)';
rho=stmat(:,1:n); rhof=rho(end,:);
% norm of dhdu
nordhdu=sqrt(sum(sum(dhdu.^2,2)*dt)) ;
% cost for current control
J=sum(sum(rho.^2+utim.^2,2)*dt)...
    +sum(rhof.^2);
% norm check for convergence
if nordhdu <= tol1
test=1 ;
disp(['nordhdu=',num2str(nordhdu)]) ;
return ;
end
```

8.6 Computer Code

```
% J1= some qty set > J. will be used in the
% iteration below to assign the updated cost
J1=J+1 ;
% step taken in terms of the maximum value of utim
tau=0.1*max(max(abs(utim))) ;
% control update procedure:
while J1>=J
% control update
utim1=utim-tau*dhdu ;
% state update
stmat=fwdstinteg(tvec,utim1,st0,b) ;
% rho and rhof extracted
rho=stmat(:,1:n); rhof=rho(end,:);
% J1=new cost
J1=sum(sum(rho.^2+utim1.^2,2)*dt)...
    +sum(rhof.^2) ;
% adjustment in the step size
tau=tau/2 ;
end
% updated control and cost
utim=utim1 ;
if abs(J1-J) <= tol2
test=1 ;
disp(['abs(J1-J) =',num2str(abs(J1-J))]) ;
end

J=J1 ;
end

function stmat=fwdstinteg(tvec,utim,st0,b)
% This routine performs the forward integration
% of the state equations using the euler step
% method.
nt=length(tvec) ; % length of the time vector
st=st0 ;
dt=tvec(2)-tvec(1) ;
% iterative procedure for the simulation
for it=1:1:(nt-1)
```

```
stmat(it,:)=st' ;
u=utim(it,:) ;
% time derivative of the states
stdot=stdyn(st,b,u) ;
% euler step method to compute the
% state at next time instant
st=st+stdot*dt ;
end
stmat(nt,:)=st ;
end

function stdot=stdyn(st,b,u)
% This routine performs the backward integration
% of the state equations using the euler step
% method.

n=length(st)/2 ; % number of sections
rho=st(1:n) ; vf=st(n+1:2*n) ;
rd(1,1)=-rho(1)*(1-rho(1))*b(1)*vf(1) ;
rd(2:n,1)=rho(1:n-1,1).*(1-rho(1:n-1,1))...
.*b(2:n).*vf(1:n-1,1)-rho(2:n,1)...
.*(1-rho(2:n,1)).*b(2:n).*vf(2:n,1) ;
vfd(1:n,1)=u ;
stdot=[rd;vfd] ;
return
end

function costmat=bkcostinteg(tvec,...
    stmat,costf,b)
% This routine computes the time derivative
% of the co-state equations
nt=length(tvec) ; % length of the time vector
cost=costf ;
dt=tvec(2)-tvec(1) ;
for it=nt:-1:2
costmat(it,:)=cost' ;
st=stmat(it,:)' ;
costdot=costdyn(cost,b,st) ;
```

8.6 Computer Code

```
    cost=cost-costdot*dt ;
  end
  costmat(1,:)=cost ;
end

function costdot=costdyn...
    (cost,b,st)
% This routine computes the time
% derivative of the co-state vector
n=length(cost)/2 ; % no of sections
rho=st(1:n); vf=st(n+1:2*n);
p=cost(1:n); pv=cost(n+1:2*n);
pd(1:n-1,1)=-(2*rho(1:n-1,1)...
+p(1:n-1,1).*((-1+2*rho(1:n-1,1))...
.*b(1:n-1).*vf(1:n-1))+p(2:n)...
.*((1-2*rho(1:n-1)).*b(2:n).*vf(1:n-1))) ;
pd(n,1)=-(2*rho(n)+p(n)*...
((-1+2*rho(n))*b(n)*vf(n))) ;
pvd(1:n-1,1)=-(p(1:n-1).*(-rho(1:n-1)...
.*(1-rho(1:n-1)).*b(1:n-1))...
+p(2:n).*(rho(1:n-1).*(1-rho(1:n-1))...
.*b(2:n))) ;
pvd(n,1)=-(p(n)*(-rho(n)*(1-rho(n))*b(n))) ;
costdot=[pd;pvd] ;
end
```

Chapter 9

Distributed Feedback Control 1-D

The goal of this chapter is on developing feedback control strategies for models that can be used for evacuation control. The first step towards this goal is to obtain mathematical models governing crowd dynamics. The development of pedestrian evacuation dynamic systems follows from the traffic flow theory in one-dimensional space. The main conservation equations used in modeling the vehicle traffic flow and the pedestrian evacuation flow are the same with the exception that vehicle traffic is a one-dimensional space problem and the evacuation system is a 2-D space problem. This chapter presents design of nonlinear feedback controllers for two different models representing evacuation dynamics in 1-D. The models presented here are based on the laws of conservation of mass and momentum. The first model is the classical one equation model for a traffic flow based on conservation of mass with a prescribed relationship between density and velocity. The other model is a two equation model in which the velocity is independent of the density. This model is based on conservation of mass and momentum. The equations of motion in both cases are described by nonlinear partial differential equations. We address the feedback control problem for both models. The objective is to synthesize a nonlinear distributed feedback controller that guarantees stability of a closed loop system. The problem of control and stability is formulated directly in the framework of partial differential equations. Sufficient conditions for Lyapunov stability for distributed control are derived.

9.1 Introduction

This chapter presents design of nonlinear feedback controllers for two models representing crowd dynamics in 1-D. The models are based on the laws of conservation of mass and momentum. The first model is the classical one-equation model for a traffic flow based on conservation of mass with a prescribed relationship between density and velocity. The model dynamics are represented by a single partial differential equation. The other model is a two equation model in which the velocity is independent of the density. This model is based on conservation of mass and momentum. As such, the model dynamics are represented by means of a system of two partial differential equations. The equations of motion in both cases are described by nonlinear partial differential equations. The system is distributed, i.e. both the state and control variables are distributed in time and space.

The control objectives are to design feedback controllers for removing people from the evacuation area effectively by generating distributed control commands. We address the feedback control problem for both models. The objective is to synthesize a nonlinear distributed feedback controller that guarantees stability of a closed loop system. The problem of control and stability is formulated directly in the framework of partial differential equations. Sufficient conditions for Lyapunov stability for distributed control are derived. There are two approaches to the design of feedback controllers for distributed systems. In the conventional approach, the distributed mathematical model is approximated by a lumped parameter model having finite dimensions. The spatial discretization of the system is performed using either the finite difference or the finite element method. The controllers are designed on the basis of the resulting linear or nonlinear ordinary differential equation model using known techniques available for such systems [22, 36]. This approach however has certain disadvantages. By neglecting the infinite dimensional nature of original system, design of controllers may result in instability even though the resulting finite dimensional system is stable using

9.2 Modeling

the same controllers. Moreover, properties like controllability and observability depend on the method of discretization used [83]. Thus, in order to avoid errors introduced by spatial discretization, it is desirable to formulate the control and stability problem directly in the framework of a distributed model of partial differential equations. In this chapter the latter approach is used.

Here we are interested in designing a feedback controller for evacuating pedestrians from a 1-D area (e.g., corridor). We adopt the method of feedback linearization for control design. The method works by canceling the nonlinearities in the system and is well known in nonlinear control for ODEs [59]. This method is introduced to quasi-linear hyperbolic PDEs. This chapter presents the feedback linearization control design for the LWR model. First we discuss the design of nonlinear feedback control for the model based on continuity equation alone and next we will discuss the feedback control design for the system described by both the equations where backstepping approach is used for the control design. In both cases the objective of control design is to synthesize a nonlinear distributed feedback controller that stabilizes the system and guarantees stability in the closed loop system.

The organization of this chapter is as follows. Section 9.2 presents the two mathematical models. In Sect. 9.3 we formulate the control model and present feedback control design for the first model. This section also studies Lyapunov stability for this model and finally presents some simulation results. The feedback control however is unbounded. In Sect. 9.4 we modify the control law to take care of the unboundedness. Section 9.5 presents the feedback control design and stability analysis for second model. Simulation results for closed loop are presented.

9.2 Modeling

In this section mathematical models of the evacuation problem are presented. We will discuss the evacuation model of a 1-D single exit corridor of length L. The model is similar to the 1-D traffic flow model. Both these models are similar to fluid flow and are based on the principle of conservation. The model is described by a nonlinear hyperbolic partial differential equation. The first model is based on continuity equation alone. We call this model One Equation Model. The second model is based on a system of PDEs. There are two partial differential equations that we use to model the control problem. The first is the equation of conservation of mass and the second is the conservation of momentum. We call this a Two Equation Model.

There are two main approaches to modeling. One approach is microscopic [48] where each individual is taken into consideration and his behavior is expressed by a set of rules or an equation involving adjacent individuals. The other approach is macroscopic [68]. Here the overall behavior of the flow of people is considered. The area is treated as a series of sections within each of which the density and average velocity of people can be measured for a given time. The changes in these variables may then be described using partial differential equations. The models presented here are macroscopic with the dynamics being represented in terms of density, flow and speed. As a result the system is distributed with all the parameters as functions of space and time.

9.2.1 One Equation Model

This model is based on equation of conservation of mass. The conservation law of mass in case of an evacuation system means that the number of people is conserved in the system. Let us consider the case of a single exit 1-D corridor of length L. Let $\rho(x,t)$ denote the density of people as a function of position vector x and time t, $q(x,t)$ the flow at a given x and time t and, and $v(x,t)$ the velocity vector field associated with the flow. The conservation of mass equation holds and is given by:

$$\frac{\partial \rho(x,t)}{\partial t} + \text{div}(q(x,t)) = 0 \tag{9.1}$$

with following initial and boundary conditions

$$\rho(x,t_0) = \rho_0(x) \tag{9.2}$$
$$\rho(0,t) = 0 \text{ and } \rho(L,t) = 0 \,\forall\, t \in [0,\infty) \tag{9.3}$$

Here $\rho(x,t) \in H^2[(0,L),\Re]$ with $H^2[(0,L),\Re]$ being the infinite dimensional Hilbert space of 1-D like vector function defined on the interval $[0,L]$ whose spatial derivatives upto second order are square

9.2 Modeling

integrable with a specified L_2 norm. $q(x,t) \in H^2[(0,L),\Re]$ and $\rho_0(x) \in H[(0,L),\Re]$. The vectors $x \in [0,L] \subset \Re$ and $t \in [0,\infty)$ denote position and time respectively. For the rest of the chapter it will be assumed that the vector spaces are Sobolev spaces [18]. The flow q is obtained as a product of density and velocity as

$$q(x,t) = \rho(x,t)v(x,t) \qquad (9.4)$$

The dynamics for 1-D are therefore given by

$$\frac{\partial \rho(x,t)}{\partial t} + \frac{\partial (\rho(x,t) v(x,t))}{\partial x} = 0 \qquad (9.5)$$

subject to the initial conditions and boundary conditions given by (9.2) and (9.3) respectively. To describe the relationship between velocity vector field $v(x,t)$ and density $\rho(x,t)$ we need one more equation. Here we make use of Greenshields model

$$v(x,t) = v_f(x,t)\left(1 - \frac{\rho(x,t)}{\rho_{\max}}\right) \qquad (9.6)$$

where $v_f(x,t)$ is the free flow speed and $\rho_{\max}$ is the jam density that is the maximum number of people that could possibly fit a single cell. Using (9.6) in (9.5) we get modified one-equation model given by

$$\frac{\partial \rho(x,t)}{\partial t} + \frac{\partial}{\partial x}\left[v_f(x,t)(1 - \frac{\rho(x,t)}{\rho_{\max}})\rho(x,t)\right] = 0 \qquad (9.7)$$

subject to conditions (9.2) and (9.3).

9.2.2 Two Equation Model

The model presented earlier is one of the original models representing the traffic flow dynamics. Since its appearance a number of other models have appeared in literature where velocity is independent of density. One such model is being presented here. In this section we consider a higher order model or more precisely a system of two partial differential equations for a 1-D corridor. This model consists of conservation of mass equation coupled with a second equation based on the principle of conservation of momentum. The first equation is the conservation of mass equation (9.5). The second equation is derived from conservation of momentum for 1-D flow [3] given by

$$\frac{\partial\left(\rho\left(x,t\right)v\left(x,t\right)\right)}{\partial t}+\frac{\partial\left(\rho\left(x,t\right)v\left(x,t\right)^{2}\right)}{\partial x}=-\frac{\partial p\left(x,t\right)}{\partial x}$$

where $p(x,t) \in H^1[(0,L), \Re]$ is pressure. The relationship between density and flow is given by (9.4). Thus we have the dynamics of an evacuation system given by following two-equation model

$$\frac{\partial \rho(x,t)}{\partial t} + \frac{\partial q(x,t)}{\partial x} = 0 \qquad (9.8)$$

$$\frac{\partial q(x,t)}{\partial t} + \frac{\partial}{\partial x}\left(\frac{q(x,t)^2}{\rho(x,t)}\right) = -\frac{\partial p}{\partial x} \qquad (9.9)$$

subject to the following initial and boundary conditions

$$\rho(x,t_0) = \rho_0(x), \, q(x,t_0) = q_0(x) \qquad (9.10)$$

and

$$\rho(0,t) = \rho(L,t) = 0$$
$$q(0,t) = q(L,t) = 0 \, \forall \, t \in [0, \infty) \qquad (9.11)$$

9.3 Feedback Control for One-Equation Model

In this section we formulate the control model and present feedback control design for the one-equation model. This section also studies Lyapunov stability and presents some simulation results.

9.3.1 Continuity Equation Control Model

To formulate the control problem based on the continuity equation model we need to choose a control variable. To do so we use Greenshields model to represent the relationship between traffic density and the velocity field which is given by (9.6) as

$$v(x,t) = v_f(x,t)\left(1 - \frac{\rho(x,t)}{\rho_{\max}}\right)$$

9.3 Feedback Control for One-Equation Model

with $v_f(x,t)$ being free flow speed and ρ_m the jam density. In this model we take free flow velocity vector field $v_f(x,t)$ as the distributed control variable denoted by $u(x,t)$. If the density at a location is zero then the speed at that location will be the free flow speed. However, with the actuation system implemented, we can tell people to change the speed. Also the traffic density affects the achievable speeds, therefore we choose $v_f(x,t)$ as the control variable, giving us the following representation

$$\frac{\partial \rho(x,t)}{\partial t} = -\frac{\partial}{\partial x}\left[[u(x,t)(1-\frac{\rho(x,t)}{\rho_m})\rho(x,t)\right] \quad (9.12)$$

where $u(x,t) \in H[(0,L),\Re]$ is the control variable.

9.3.2 State Feedback Control

Here we address the problem of synthesizing a distributed state feedback controller $u(x,t)$ that stabilizes origin $\rho(x,t) = 0$ of system (9.12) or in other words control the evacuation of pedestrians from a 1-D area. We shall use the method of feedback linearization for PDEs. The method of feedback linearization works by canceling the nonlinearities in the system and is discussed for nonlinear ODEs in [59]. More specifically we consider control law of the form

$$u(x,t) = F(\rho(x,t))$$

which makes origin of the closed loop dynamics exponentially stable. Here $F(\rho)$ is a nonlinear operator mapping $H^2[(0,L),\Re]$ into $H^1[(0,L),\Re]$. Designing the operator $F(\rho)$ we get our state feedback controller as

$$F(\rho(x,t)) = -\left[(1-\frac{\rho(x,t)}{\rho_m})\rho(x,t)\right]^{-1} D\frac{\partial \rho(x,t)}{\partial x} \quad (9.13)$$

we get the following closed loop dynamics

$$\frac{\partial \rho(x,t)}{\partial t} - D\frac{\partial^2 \rho(x,t)}{\partial x^2} = 0 \quad (9.14)$$

with boundary conditions given by (9.2) and (9.3). The closed loop dynamics represent the heat equation with D being the diffusion constant. Equation (9.14) suggests that the people are diffusing or in other words the motion of people is because of diffusion only. The rate of diffusion is determined by the diffusion constant D.

9.3.3 Lyapunov Stability Analysis

In order to check the stability of the closed loop system (9.14) we use the Lyapunov function analysis. The stability problem is to establish sufficient conditions for which the origin of the closed loop dynamics (9.14) is exponentially or asymptotically stable. Within the framework of our system the definition of Lyapunov stability can be established by writing system dynamics (9.14) as an abstract differential equation in terms of operator theory. Towards that end, we introduce a differential operator A on $H^2[(0,L), \Re]$ defined by

$$A\varphi = k\frac{\partial^2 \varphi}{\partial x^2} \; \forall \; \varphi \in D(A) \qquad (9.15)$$

Here $D(A) \subset H^2[(0,L), \Re]$ is the domain of operator A defined as

$$D(A) = \{\varphi \in H^2 : \varphi, \varphi' \in H^2[0,L]; \varphi(0) = \varphi(L) = 0\}$$

The system dynamics can be written using operator (9.15) which makes the dynamics look like ordinary differential equation in Sobolev (Banach) spaces [8, 9]. Thus the abstract version or state space representation of (9.14) can be put into the form

$$\frac{d}{dt}\rho(t) = A\rho(t) \; \forall \; t > 0; \rho(0) = \rho_0 \qquad (9.16)$$

The operator A is known to generate a strongly continuous semigroup $U(t)$ of bounded linear operators on a normed linear space $H^2[(0,L), \Re]$. The system motion starting from any initial state $\rho(t_0)$ at time t_0 is defined by $U(t)\rho(t_0)$. Thus a partial differential equation can be regarded as an evolution system where $U(t)$ evolves ρ_0 forward in time. The stability problem is to establish sufficient conditionsfor which the origin of the closed loop dynamics (9.14) is exponentially stable. Within the framework of our system the definition of stability in terms of Lyapunov can be established as follows

Definition 9.3.1 An equilibrium state ρ_{eq} of a dynamical system (9.14) is stable with respect to a specified norm $\|\rho(x,t)\|$ if for every real number $\varepsilon > 0$ there exists a real number $\delta(\varepsilon, t_0) > 0$ such that

$$\|\rho(t_0) - \rho_{eq}\| < \delta \Rightarrow \|U(t)\rho(t_0) - \rho_{eq}\| < \varepsilon \; \forall \; t > t_0$$

9.3 Feedback Control for One-Equation Model

If in addition $\|U(t)\rho(t_0) - \rho_{eq}\| \to 0$ as $t \to \infty$ then the equilibrium is said to be asymptotically stable. Furthermore, if there exist two positive constants a and b such that

$$\|U(t)\rho(t_0) - \rho_{eq}\| \le a \|\rho(t_0) - \rho_{eq}\| e^{-b(t-t_0)} \ \forall \ t > t_0$$

is satisfied, then ρ_{eq} is said to be exponentially asymptotically stable. More precisely the problem is to find a condition under which operator A generates an exponentially stable semigroup $U(t)$ that satisfies the following growth property with respect to a specified norm

$$\|U(t)\|_2 \le a e^{-bt} t \ge 0 \tag{9.17}$$

where $b > 0$. In other words we say that A generates an exponentially stable semigroup $U(t)$ [28]. The determination of conditions for which estimate (9.17) is satisfied amounts to establishing conditions for which the null state of the linear system (9.14) is exponentially asymptotically stable with respect to the specified norm i.e. $\|\rho(t)\|_2 \to 0$ as $t \to \infty$. It should be noted here that for infinite dimensional systems stability with respect to one norm does not necessarily imply stability with respect to others unlike finite dimensional systems where all norms are equivalent. For our system we have chosen the following L_2 norm defined by

$$\|\rho(t,x)\|_2 \to \left[\int_0^L |\rho(t,x)|^2 \, dx\right]^{1/2} \tag{9.18}$$

This norm represents the total energy of the system at any time. Thus the exponential for this norm implies that the systems total energy goes to zero as time goes to infinity.

Let us consider a Lyapunov functional $V(t)$ for the system (9.14). Here $V : H^2[0, L] \to \Re_+$ is a smooth functional of the form

$$V(t) = \tfrac{1}{2} \|\rho(t,x)\|_2^2 = \tfrac{1}{2} \int_0^L |\rho(t,x)|^2 dx \tag{9.19}$$

Using the norm properties we can easily see that $V(t)$ is a positive definite function. The time rate of change of $V(t)$ using Leibniz rule in (9.19) is given as

$$\frac{dV(t)}{dt} = \int_0^L \rho(x,t) \frac{\partial \rho(x,t)}{\partial t} dx \tag{9.20}$$

Using (9.12) and integrating (9.20) we get

$$\frac{dV(t)}{dt} = \left[k\rho(x,t)\frac{\partial \rho(x,t)}{\partial x}\right]_0^L - k\int_0^L \left(\frac{\partial \rho(x,t)}{\partial x}\right)^2 dx$$

The first term vanishes by boundary condition (9.3). For the second integral we make use of Gagliardo-Nirenberg-Sobolev Inequality [25]. The inequality as applied to our case states

$$\|\rho(x,t)\|_2 \leq C \|\nabla \rho(x,t)\|_2 \qquad (9.21)$$

where C is a positive real number. Using (9.21) we have

$$\int_0^L \left(\frac{\partial \rho(x,t)}{\partial x}\right)^2 dx \geq C^{-2} \int_0^L \rho^2(t,x) dx$$

The rate of change of $V(t)$ can be thus be bounded by

$$\frac{dV(t)}{dt} \leq -kC^{-2}\int_0^L \rho^2(t,x)dx \leq -2kC^{-2}V(t) = -\beta V(t)$$

It follows that

$$V(t) \leq V(t_0)e^{-\beta(t-t_0)}$$

or

$$\|\rho(t,x)\|_2 \leq \|\rho(t_0,x)\|_2 \, e^{-\beta(t-t_0)}$$

with $\beta = 2kC^{-2}$. As long as $\beta > 0$, null state of (9.14) is exponentially stable and condition (9.15), $\|U(t)\|_2 \leq e^{-b(t-t_0)}$ is satisfied with $b = \beta$. Thus equilibrium of closed loop system (9.14) using feedback control (9.13) is exponentially stable.

9.3.4 Simulation Results

This section shows simulation results for closed loop system (9.14) using the control law (9.13). The numerical method used is the Lax-Friedrichs scheme [67]. For simulation the initial distribution for density is considered to be Gaussian. The initial density distribution is given by.

$$\rho(x,0) = G\exp(-(x-a)^2)$$

9.3 Feedback Control for One-Equation Model

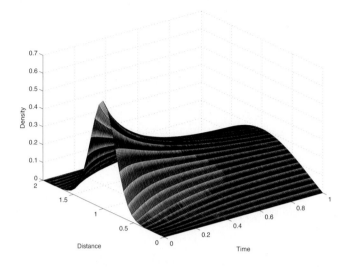

Fig. 9.1. Density response for one-equation model

with a being the centre of the Gaussian distribution and G is the highest magnitude of the distribution. We have three simulation results for the one-equation model. The control action response for a corridor with exit at the right is shown in Fig. 9.1. The simulation

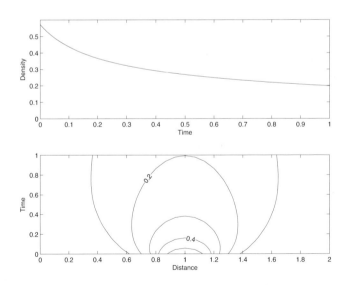

Fig. 9.2. Contours of the density response for one equation model

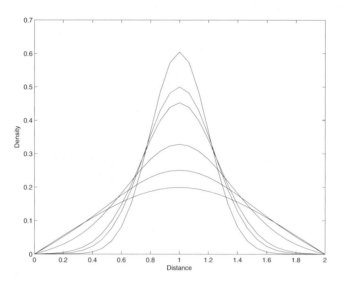

Fig. 9.3. Density response at different time instants for one-equation model

shows the density response as a mesh plot. After some finite time the density has decreased to zero. The second simulation is shown as a contour plot in Fig. 9.2 where the contour lines vary from 0.6 to 0.1 and as seen from plots density at every point in space is decreasing exponentially with time. The third simulation shows the density response at different time instants in Fig. 9.3, where the density flattens out with time. The plots indicate the diffusion of people throughout the length of the corridor.

9.4 Control Saturation

The feedback controller for the one-equation model discussed in previous section have saturation issues which will be discussed here. The control values in the previous controller become unbounded. So far we have assumed unbounded control. However, the unboundedness poses a limitation on the practical implementation of the controller. Therefore in order to take care of control saturation we redesign our controller so that there is no unboundedness of control. In the control model (9.12) by using the control law (9.14) we have the issue of control saturation. The controllaw (9.14) is

9.4 Control Saturation

$$F(\rho(x,t)) = -\left[(1 - \frac{\rho(x,t)}{\rho_m})\rho(x,t)\right]^{-1} k\frac{\partial \rho(x,t)}{\partial x}$$

With this control we could cancel the nonlinearity in the system and convert the system into a heat equation. But the issue with this control is that it is unbounded because of the inverse. At $\rho(x,t) = \rho_m$ the control becomes unbounded. This kind of situation is practically impossible as we have limitations on the control variable. So in order to take care of control saturation we redesign our controllers so that there is no unboundedness of control. Now we choose the following control law

$$F(\rho(x,t)) = \begin{cases} \frac{1}{\rho} D \frac{\partial \rho}{\partial x} & \rho \neq 0 \\ v_{min} & \rho = 0 \end{cases} \quad (9.22)$$

With this control law we get the following closed loop dynamics

$$\frac{\partial \rho(x,t)}{\partial t} = -\frac{\partial}{\partial x}\left[(1 - \frac{\rho(x,t)}{\rho_m})D\frac{\partial \rho(x,t)}{\partial x}\right]$$

or

$$\frac{\partial \rho(x,t)}{\partial t} + \left(1 - \frac{\rho(x,t)}{\rho_m}\right)\frac{\partial^2 \rho(x,t)}{\partial x^2} + \frac{1}{\rho_m}\left(\frac{\partial \rho(x,t)}{\partial x}\right)^2 = 0 \quad (9.23)$$

Let us consider a Lyapunov functional $V(t)$ for the system (9.23) given by (9.19) as:

$$V(t) = \frac{1}{2}\|\rho(t,x)\|_2^2 = \frac{1}{2}\int_0^L |\rho(t,x)|^2 dx$$

The time rate of change of $V(t)$ using Leibniz rule is given as

$$\frac{dV(t)}{dt} = \int_0^L \rho(x,t)\frac{\partial \rho(x,t)}{\partial t} dx$$

Using (9.23) we get

$$\frac{dV(t)}{dt} = \int_0^L -\rho(x,t)\left[\left(1 - \frac{\rho(x,t)}{\rho_m}\right)\frac{\partial^2 \rho(x,t)}{\partial x^2} + \frac{1}{\rho_m}\left(\frac{\partial \rho(x,t)}{\partial x}\right)\right]dx$$

or

$$\frac{dV(t)}{dt} = \int_0^L -\rho(x,t)\left(1 - \frac{\rho(x,t)}{\rho_m}\right)\frac{\partial^2 \rho(x,t)}{\partial x^2} dx$$

$$+ \int_0^L \frac{\rho(x,t)}{\rho_m}\left(\frac{\partial \rho(x,t)}{\partial x}\right)^2 dx \quad (9.24)$$

Equation (9.24) can be written as

$$\frac{dV(t)}{dt} = -I_1 + I_2$$

Integrating the first integral gives us

$$I_1 = \rho(x,t)(1 - \frac{\rho(x,t)}{\rho_m})\frac{\partial\rho(x,t)}{\partial x}\Big|_0^L$$
$$- \int_0^L (1 + \frac{2\rho(x,t)}{\rho_m})\left(\frac{\partial\rho(x,t)}{\partial x}\right)^2 dx \quad (9.25)$$

The first term of (9.25) vanishes by boundary condition (9.3). Thus (9.24) reduces to the following

$$\frac{dV(t)}{dt} = -\int_0^L (1 + \frac{\rho(x,t)}{\rho_m})\left(\frac{\partial\rho(x,t)}{\partial x}\right)^2 dx$$

or

$$\frac{dV(t)}{dt} = -\int_0^L \left(\frac{\partial\rho(x,t)}{\partial x}\right)^2 dx$$
$$- \frac{1}{\rho_m}\int_0^L \rho(x,t)\left(\frac{\partial\rho(x,t)}{\partial x}\right)^2 dx \quad (9.26)$$

For the first integral of (9.26) we make use of Gagliardo-Nirenberg-Sobolev Inequality (9.21) which for our case is

$$\|\rho(x,t)\|_2 \leq C \|\nabla\rho(x,t)\|_2$$

where C is a positive real number. Using (9.21) we have

$$\int_0^L \left(\frac{\partial\rho(x,t)}{\partial x}\right)^2 dx \geq C^{-2} \int_0^L \rho^2(t,x) dx$$

The second integral of (9.26) can be written as

$$I = \int_0^L \rho(x,t)\left(\frac{\partial\rho(x,t)}{\partial x}\right)^2 dx$$

9.5 Feedback Control for Two Equation Model

Using Young's Inequality [25] we can show that

$$I \leq \frac{1}{2}\int_0^L \rho^2(t,x)\mathrm{d}x + \int_0^L \left(\frac{\partial \rho(x,t)}{\partial x}\right)^4 \mathrm{d}x$$

The rate of change of $V(t)$ can be thus be bounded by

$$\frac{\mathrm{d}V(t)}{\mathrm{d}t} \leq -kC^{-2}\int_0^L \rho^2(t,x)\mathrm{d}x - \frac{1}{2\rho_m}\int_0^L \rho^2(t,x)\mathrm{d}x$$
$$- \frac{1}{2\rho_m}\int_0^L \left(\frac{\partial \rho(x,t)}{\partial x}\right)^4 \mathrm{d}x$$

or

$$\frac{\mathrm{d}V(t)}{\mathrm{d}t} \leq -\alpha \int_0^L \rho^2(t,x)\mathrm{d}x - \beta \int_0^L \rho^4(t,x)\mathrm{d}x \tag{9.27}$$

where $\alpha = kC^{-2} - 1/2\rho_m$. Using Young's inequality again to the second term of (9.27) we get the following

$$\frac{\mathrm{d}V(t)}{\mathrm{d}t} \leq -\alpha \int_0^L \rho^2(t,x)\mathrm{d}x - \beta \int_0^L \rho^4(t,x)\mathrm{d}x$$

where $\beta = k^{-2}/2\rho_m$ with k being a positive constant. Both the first and second terms are positive definite functions, therefore as long as $\alpha > 0$ and $\beta > 0$ we have

$$\frac{\mathrm{d}V(t)}{\mathrm{d}t} < 0$$

It follows that null state of (9.22) is asymptotically stable using feedback control (9.22).

9.5 Feedback Control for Two Equation Model

In this section we design a feedback control for the two-equation model of the evacuation system given by (9.8) and (9.9) subject to boundary conditions (9.10) and (9.11) using backstepping approach. The Lyapunov functional (9.19) which was used as a stability analysis tool for one-equation model will be used as a feedback control design tool for this system. The design of feedback control is done in such a way that Lyapunov functional or its derivative has certain properties that guarantee boundedness or convergence to an equilibrium point.

9.5.1 Two Equation Control Model

In the two equation model (9.8) and (9.9) we choose divergence of pressure $\partial p(x,t)/\partial x$ as distributed control variable $u(x,t)$ which gives us the following control model

$$\frac{\partial \rho(x,t)}{\partial t} = -\frac{\partial q(x,t)}{\partial x}$$

$$\frac{\partial q(x,t)}{\partial t} = -\frac{\partial}{\partial x}\left(\frac{q(x,t)^2}{\rho(x,t)}\right) + u(x,t)$$

This system can be rewritten in the form

$$\frac{\partial \rho(x,t)}{\partial t} = -\frac{\partial q(x,t)}{\partial x} \qquad (9.28)$$

$$\frac{\partial q(x,t)}{\partial t} = \bar{u}(x,t) \qquad (9.29)$$

where $\bar{u}(x,t) = -\frac{\partial}{\partial x}\left(\frac{q(x,t)^2}{\rho(x,t)}\right) + u(x,t)$.

9.5.2 State Feedback Control Using Backstepping

Here we address the problem of synthesizing a distributed state feedback controller $\bar{u}(x,t)$ that stabilizes origin $\rho(x,t) = 0, q(x,t) = 0$ of control system (9.28) and (9.29). More specifically we consider control law

$$\bar{u}(x,t) = \bar{F}\left(\rho(x,t), q(x,t)\right)$$

such that origin of closed loop dynamics are exponentially stable. $\bar{F}$ is a nonlinear operator mapping $H^2[(0,L), \Re]$ into $H^1[(0,L), \Re]$. The control strategy adopted here is similar in principle to feedback control by backstepping for ordinary differential equations [59] and is extended to PDEs. Backstepping is a Lyapunov-based control method of feedback linearization. It is a recursive method that designs the feedback control law based on the choice of the Lyapunov function. It breaks the design problem for a system of equations into a sequence of design problems for scalar systems. We proceed with the control design as follows:

1. First we design control law for (9.28) where $q(x,t)$ can be viewed as an input. We proceed to design a conceptual control

9.5 Feedback Control for Two Equation Model

law $q(x,t) = G(\rho(x,t))$ to stabilize origin $\rho(x,t) = 0$. G is a nonlinear operator mapping $H^2[(0,L), \Re]$ into $H^2[(0,L), \Re]$. With control law

$$q(x,t) = G(\rho(x,t)) = -\int_0^x \frac{\partial^2 \rho(x,t)}{\partial m^2} dm \qquad (9.30)$$

we get the conceptual closed-loop system for (9.28) as

$$\frac{\partial \rho(x,t)}{\partial t} = \frac{\partial^2 \rho(x,t)}{\partial x^2} \qquad (9.31)$$

which is similar to (9.14). As we have already shown the origin of this equation is asymptotically exponentially stable with a conceptual Lyapunov functional (9.19)

$$V(t) = \tfrac{1}{2} \|\rho(t,x)\|_2^2 = \tfrac{1}{2} \int_0^L |\rho(t,x)|^2 dx$$

for this system which satisfies $dV(t)/dt \leq -\beta V(t)$ and ensures the stability of above equation.

2. We have used the term "conceptual" with the control law, closed loop system and the Lyapunov function. This is done in order to stress the fact that the control law cannot be implemented in practice as $q(x,t)$ is not a control variable. However, this conceptual design helps us to recognize the benefit of the input $q(x,t)$ being close to $G(\rho)$. From the knowledge of the conceptual Lyapunov function $V(t)$ we want to design a smooth feedback control for stabilizing the origin of the overall system. We therefore add to the conceptual Lyapunov function (9.19) a term penalizing the difference between q and $G(\rho)$. For this purpose we rewrite the dynamics (9.28) as

$$\frac{\partial \rho(x,t)}{\partial t} = -\frac{\partial G(\rho(x,t))}{\partial x} - \frac{\partial}{\partial x}(q(x,t) - G(\rho(x,t)))$$

Defining the difference between q and $G(\rho)$ by an error a new variable $z(x,t) = q(x,t) - G(\rho(x,t))$ we get the following dynamics.

$$\frac{\partial \rho(x,t)}{\partial t} = -\frac{\partial G(\rho(x,t))}{\partial x} - \frac{\partial z(x,t)}{\partial x} \qquad (9.32)$$

$$\frac{\partial z(x,t)}{\partial t} = u_n(x,t) \qquad (9.33)$$

where $u_n(x,t) = \bar{u}(x,t) - \partial G(\rho(x,t))/\partial t$ is the new control variable and $z(x,t) \in H^2[(0,L), \Re]$.

3. Now let us modify the Lyapunov functional (9.19) by adding an error term to it thus resulting in the Lyapunov function for the overall system as

$$\begin{aligned} V_a(t) &= V(t) + \tfrac{1}{2}\|z(x,t)\|_2^2 \\ &= \tfrac{1}{2}\int_0^L |\rho(x,t)|^2\,dx + \tfrac{1}{2}\int_0^L |z(x,t)|^2\,dx \end{aligned} \quad (9.34)$$

The time rate of change of this functional using (9.32) and (9.33) is given as

$$\frac{dV_a(t)}{dt} = \int_0^L \rho(x,t)\frac{\partial^2 \rho(x,t)}{\partial x^2}dx + \int_0^L \rho(x,t)\frac{\partial z(x,t)}{\partial x}dx \\ + \int_0^L z(x,t) u_n(x,t)\,dx$$

Using (9.21) we know that the first term is bounded by $-\beta V(t)$. Therefore

$$\frac{dV_a(t)}{dt} \leq -\beta V(t) + \int_0^L \rho(x,t)\frac{\partial z(x,t)}{\partial x}dx + \int_0^L z(x,t) u_n(x,t)\,dx$$

We have to choose a new control law $u_n(x,t)$ in such a manner that the time derivative of new functional or the sum of second and third terms is also bounded by a negative definite function. By choosing the following control law

$$u_n(x,t) = -\hat{k}z(x,t) - z(x,t)^{-1}\rho(x,t)\frac{\partial z(x,t)}{\partial x} \quad (9.35)$$

we get the following result

$$\frac{dV_a(t)}{dt} \leq -\beta V(t) - \hat{k}\|z(x,t)\|_2^2 = -2\beta V_a(t)$$

with $\hat{k} = 2\beta > 0$. This shows that the origin of closed loop system (9.28) and (9.29) is exponential stable.

9.5 Feedback Control for Two Equation Model

The control law is given by the partial differential-integral equation

$$u(x,t) = u_n(x,t) + \frac{\partial G(\rho(x,t))}{\partial t} + \frac{\partial}{\partial x}\left(\frac{q(x,t)^2}{\rho(x,t)}\right) \qquad (9.36)$$

with $u_n(x,t)$ given by (9.35) as

$$u_n(x,t) = -\widehat{k}z(x,t) - z(x,t)^{-1}\rho(x,t)\frac{\partial z(x,t)}{\partial x}$$

where $z(x,t) = q(x,t) - G(\rho(x,t))$. Hence the closed loop dynamics for two equation model are exponentially stable.

9.5.3 Simulation

In this section we show simulation results for the closed loop system (9.28) and (9.29) using controller (9.36). We have used the same Lax-Friedrichs numerical technique as before. For simulation the initial distribution for both density and flow is considered to be Gaussian. The simulation results are shown in the following figures. The density plots are shown in Figs. 9.4–9.6. The flow plots are shown in Fig. 9.7 and Fig. 9.8. As is seen from the plots flow of people at every point in space is decreasing exponentially with time.

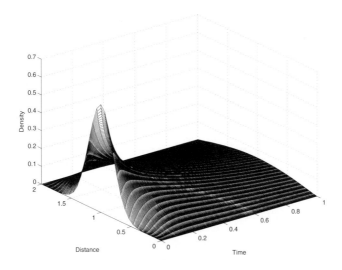

Fig. 9.4. Density response for the two-equation model

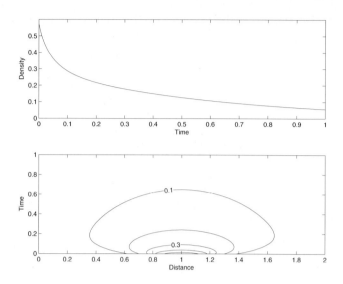

Fig. 9.5. Contours of the density response for two-equation model

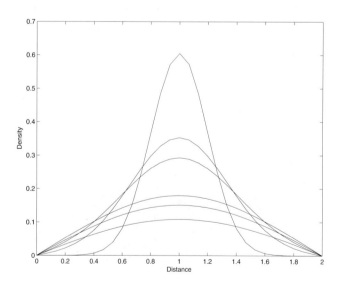

Fig. 9.6. Density response at different time instants for two-equation model

9.5 Feedback Control for Two Equation Model

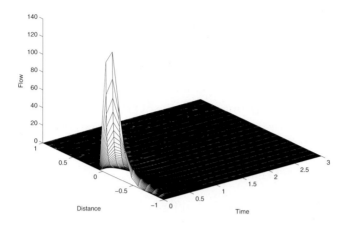

Fig. 9.7. Flow response for the two-equation model

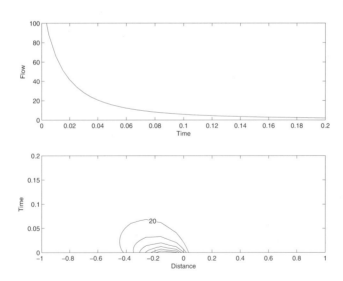

Fig. 9.8. Contours of flow response for two-equation model

9.6 Exercises

Problem 9.6.1 Add the term $1/\rho_m \partial \rho(x,t)/\partial x$ to control law (9.22) and repeat the Lyapunov Analysis. Is the system still exponentially stable?

Problem 9.6.2 Work out the details of Sobolev inequality for (9.20).

Problem 9.6.3 Work out the details of Young's inequality for (9.26).

Problem 9.6.4 What is Leibnitz rule?

Problem 9.6.5 What is the difference between a Banach and a Sobolev space?

Problem 9.6.6 Prove that the norms are not equivalent in infinite dimensional case.

Problem 9.6.7 Prove that the norms are not equivalent in infinite dimensional case.

9.7 Computer Code

In this section the code used for the above simulations is included. The code is in matlab and consists of m-file named feedback_1d_mass and feedback_1d_momentum.

9.7.1 feedback_1d_mass

```
%n=10;   %points in x direction
n=20;
%m=300;  %time samples
m=600;
T=3;
L=2;
h=L/(n-1); % spacing in x-direction
k=T/(m-1); % spacing in time
s=k/(h^2);
x=[-1:h:1];
t=[0:k:3];
rm=1000;
```

9.7 Computer Code

```
std=L/8;meanx=0; Row_0 = 100*exp(-(x-meanx).^2./...
(2*std^2)) / (std*sqrt(2*pi));
Row=Row_0;
Row(1:m,1)=0;
for r=1:m-1
    for c=2:n-1
    Row(r+1,c)=Row(r,c)+s*(Row(r,c+1)-.........
    2*Row(r,c)+Row(r,c-1));
    end
end
for i=1:m
    for j=1:n-1
del_row(i,j)=1/h*(Row(i,j+1)-Row(i,j));
 end
end
del_row(m,n)=0;

for i=1:m
    del2_row(i,1)=1/(h)^2*(Row(i,2)-2*Row(1,1)+0);
    for j=2:n-1
del2_row(i,j)=1/h^2*(Row(i,j+1)-2*Row(i,j)+Row(i,j-1) );
 end
end
del2_row(m,n)=0;
A=1./(1-2*Row/rm);
B=(del_row);
C=(del2_row);

vf=(A/B)*C;
v=vf.*(1-Row/rm);
 for i=1:m
  for j=1:n
    if v(i,j)<0
    v(i,j)=0;
    end
  end
end
q=Row.*v;
```

```
figure(1)
mesh(t,x,Row')
% title('Heat Equation solution')
xlabel('Time')
ylabel('Distance')
zlabel('Density')
figure(2)
subplot(2,1,1),
plot(t',Row(:,n/2));
axis([0 1.5 0 200])
% title('Density plot at x=L/2')
xlabel('Time')
ylabel('Density')
subplot(2,1,2),
axis([-1 1 0 1])
[c,h1] = contour(x,t,Row); clabel(c,h1,'manual');
axis([-1 1 0 1.5])
% title('Density Contours')
xlabel('Distance')
ylabel('Time')
figure(3)
mesh(t,x,q')
% title('Heat Equation solution')
xlabel('Time')
ylabel('Distance')
zlabel('Flow')

figure(4)
subplot(2,1,1),
plot(t',q(:,n/2));
axis([0 0.2 0 100])
% title('Flow plot at x=L/2')
xlabel('Time')
ylabel('Flow')
subplot(2,1,2),
[c,h1] = contour(x,t,q); clabel(c,h1,'manual');
% title('Flow Contours')
axis([-1 1 0 0.2])
```

```
xlabel('Distance')
ylabel('Time')
```

9.7.2 feedback_1d_momentum

```
n=20;%points in x dir
m=800;  %time samples
T=3;
L=2;
delta_x=L/(n-1);
delta_t=T/(m-1);
x=[-1:delta_x:1];
t=[0:delta_t:3];
vm=0.1;
row_m=1;
lamda=delta_t/delta_x;

std=L/8;meanx=0; row_0 = 10*exp(-(x-meanx)..........
^2./(2*std^2)) / (std*sqrt(2*pi));
row=row_0;
row(1:m,1)=0;

std=L/8;meanx=0; q_0 = 10*exp(-(x-meanx).........
^2./(2*std^2)) / (std*sqrt(2*pi));
q=q_0;
q(1:m,1)=0;
for i=1:m-1
    for j=2:n-1
        f1(i,j+1)=1/delta_x*(row(i,j+1)-row(i,j));
        f1(i,j-1)=1/delta_x*(row(i,j)-row(i,j-1));
        row(i+1,j)=0.5*(row(i,j+1)+row(i,j-1)).........
        -lamda/2*(f1(i,j+1)-f1(i,j-1));
    end
end

for i=1:m
    del2_row(i,1)=1/(delta_x)^2*(row(i,2).......
    -2*row(1,1)+0);
    for j=2:n-1
```

```
del2_row(i,j)=1/(delta_x)^2*(row(i,j+1)........
-2*row(i,j)+row(i,j-1) );
 end
end
del2_row(m,n)=0;

for i=1:m
   F_row(i,1)=del2_row(i,1);
    F=0;
    for j=2:n
      F_row(i,j)=F+delta_x/2*(del2_row(i,j)....
      +del2_row(i,j-1));
      F=F_row(i,j);
    end
end

for i=1:m-1
    for j=1:n-1
delt_F(i,j)=1/delta_t*(F_row(i+1,j)-F_row(i,j));
 end
end
delt_F(m,n)=0;

 k=0.5;
 for i=1:m-1
         q_hat(i,1)=q(i,1)-F_row(i,1);
         q_hat(i,2)=q(i,2)-F_row(i,2);
         delx_q_hat(i,1)=1/delta_x*
            (q_hat(i,2)-q_hat(i,1));
         u(i,1)=delt_F(i,1)-k*q_hat(i,1)-.........
         (1/q_hat(i,1))*row(i,1)*delx_q_hat(i,1);
         P(i,1)=u(i,1);
         p=0;
for j=2:n-1
         q_hat(i,j)=q(i,j)-F_row(i,j);
         q_hat(i,j+1)=q(i,j+1)-F_row(i,j+1);
         if j>=n-1
         q_hat(i,j+2)=0;
         else
```

9.7 Computer Code

```
                q_hat(i,j+2) = q(i,j+2)-F_row(i,j+2);
                  end
                delx_q_hat(i,j)
                  = 1/delta_x*(q_hat(i,j+1)-q_hat(i,j));
                delx_q_hat(i,j+1)
                  = 1/delta_x*(q_hat(i,j+2)-q_hat(i,j+1));

                u(i,j)=delt_F(i,j)-k*q_hat(i,j)-......
                (1/q_hat(i,j))*row(i,j)*delx_q_hat(i,j);
                u(i,j+1)=delt_F(i,j+1)-k*q_hat(i,j+1).....
                -(1/q_hat(i,j+1))*row(i,j+1)*delx_q_hat(i,j+1);

                P(i,j)=p+delta_x/2*(u(i,j)+u(i,j-1));
                P(i,j+1)=P(i,j)+delta_x/2*(u(i,j+1)+u(i,j));
                p=P(i,j);

                f2(i,j+1)=(q(i,j+1))^2+P(i,j+1);
                f2(i,j-1)=(q(i,j-1))^2+P(i,j-1);
                q(i+1,j)=0.5*(q(i,j+1)+q(i,j-1)).....
                -lamda/2*(f2(i,j+1)-f2(i,j-1));
        end
end

figure(1)
mesh(t,x,row')
title('Backstepping solution')
xlabel('Time')
ylabel('Distance')
zlabel('Density')
figure(2)
subplot(2,1,1),
plot(t',row(:,n/2));
title('Density plot at x=L/2')
xlabel('Time')
ylabel('Density')
subplot(2,1,2),
[c,h1] = contour(t,x,row'); clabel(c,h1,'manual');
title('Density Contours')
xlabel('Disatance')
ylabel('Time')
```

Chapter 10

Distributed Feedback Control 2-D

This chapter presents design of nonlinear feedback controllers for two different models representing evacuation dynamics in 2-D. The models presented here are similar to the ones discussed in Chap. 9 but take care of the pedestrian flow in 2-D. We address the feedback control problem for both models and the sufficient conditions for Lyapunov stability for distributed control are derived.

10.1 Introduction

This chapter presents the design of nonlinear feedback controllers for two models representing evacuation dynamics in 2-D. The models presented here are based on the laws of conservation of mass and momentum. The first model is a one-equation model based on conservation of mass with a prescribed relationship between density and velocity. The model dynamics are represented by means of a single partial differential equation. The other is a two-equation model in which the velocity is independent of density. This model is based on conservation of mass and momentum. Here the dynamics are represented by means of a set of three partial differential equations. The equations of motion in both cases are described by nonlinear partial differential equations. The system is distributed, i.e. both the state and control variables are distributed in time and space.

We address the feedback control problem for both models. The objective is to synthesize a nonlinear distributed feedback controller that guarantees stability of a closed loop system. The problem of control and stability is formulated directly in the framework of partial differential equations. Sufficient conditions for Lyapunov stability for distributed control are derived.

We are interested in designing feedback controllers to evacuate pedestrians from a 2-D area. We use the same methods of feedback control as in Chap. 9 and extend the results to a 2-D case. The method of feedback linearization is used for the one-equation model which works by canceling nonlinearities in the system. For a two-equation system model the feedback control design is done by the backstepping approach. In both cases objective of control design is to synthesize a nonlinear distributed feedback controller that stabilizes the system and guarantees stability in closed loop system.

The objective of this chapter is to design feedback controllers and study their stability properties for an evacuation control system in 2-D. The dynamics that are used to model an evacuation system in this chapter are based on traffic flow theory [57] modified for the bidirectional pedestrian flow. In this chapter again we are adopting the approach of designing controllers in distributed setting. Sufficient conditions for Lyapunov stability for distributed control are also derived. First we discuss the design of nonlinear feedback control for the model based on continuity equation alone. To increase resolution and accuracy of the model we add the equation of conservation of momentum. We discuss the feedback control design for the system described by both equations. Backstepping approach for control design used in Chap. 9 will be modified for the 2-D case. In both cases the objective of control design is to synthesize a nonlinear distributed feedback controller that stabilizes the system and guarantees stability in closed loop system.

The organization of this chapter is as follows. In Sect. 10.2 we formulate the mathematical and control models and present feedback control design for the first model. This section also studies Lyapunov stability for this model. Section 10.3 presents the feedback control design and stability analysis for second model. Finally some simulation results for closed loop are presented.

10.2 Feedback Control of One-Equation Model

In this section we formulate the control model and present feedback control design for the one-equation model. This section also studies Lyapunov stability for this model and presents simulation results.

10.2.1 One-Equation Model

In this section mathematical and control models and control design for first model of the evacuation problem are presented. We will discuss the evacuation model of a 2-D single exit corridor of dimensions $L \times L$. The model is described by nonlinear hyperbolic partial differential equation.

This model is based on equation of conservation of mass. The conservation law of mass in case of an evacuation system means that the number of people is conserved in the system. Let us consider the case of a single exit 2-D corridor of dimension $L \times L$. Let $\Omega = (0, L) \times (0, L)$ be a bounded, open subset of a 2-D Euclidean space R^2 and $\partial\Omega$ be its boundary. Let $\rho(x,t)$ denote the density of people as a function of position vector x and time t. The vector $x \in \Omega \subset \Re^2$ and is expressed in terms of its coordinates as $x = [x_1, x_2]^T$. Let $q(x,t)$ be the flow at a given x and t with $q_1(x,t)$ and $q_2(x,t)$ as flow in x_1 and x_2 directions. $v(x,t)$ is the velocity vector field associated with the flow with $v_1(x,t)$ and $v_2(x,t)$ as x_1 and x_2 components respectively. The conservation of mass equation holds and is given by:

$$\frac{\partial \rho(x,t)}{\partial t} + \text{div}(q(x,t)) = 0 \qquad (10.1)$$

with initial and a boundary condition given as

$$\rho(x, t_0) = \rho_0(x) \qquad (10.2)$$
$$\rho(x, t) = 0 \; \forall \; t \in [0, \infty), \; x \in \partial\Omega \qquad (10.3)$$

Here $\rho(x,t) \in H^2[\Omega, \Re]$ with $H^2[\Omega, \Re]$ being the infinite dimensional Hilbert space of 2-D like vector function defined on domain Ω, whose spatial derivatives upto second order are square integrable with a specified L_2 norm. $q(x,t) \in H^2[\Omega, \Re]$ and $\rho_0(x) \in H[\Omega, \Re]$. The vectors $x \in \Omega \subset \Re^2$ and $t \in [0, \infty)$ denote position and time respectively. For the rest of the chapter it will be assumed that the vector spaces are Sobolev spaces [18]. The flow $q(x,t)$ is obtained as a product density and velocity as

$$q_i(x,t) = \rho(x,t)v_i(x,t), i = 1, 2 \tag{10.4}$$

The dynamics for 2-D are therefore given by

$$\frac{\partial \rho(x,t)}{\partial t} + \frac{\partial \left(\rho(x,t) v_1(x,t)\right)}{\partial x_1} + \frac{\partial \left(\rho(x,t) v_2(x,t)\right)}{\partial x_2} = 0 \tag{10.5}$$

subject to the initial conditions and boundary conditions given by (10.2) and (10.3) respectively. To describe the relationship between velocity vector field $v(x,t)$ and density $\rho(x,t)$ we again make here use of Greenshields model in 2-D

$$v_i(x,t) = v_{if}(x,t)\left(1 - \frac{\rho(x,t)}{\rho_{\max}}\right) \tag{10.6}$$

where $v_{if}(x,t)$ is the free flow speed in x_i direction $i = 1, 2$ and $\rho_{\max}$ is the jam density that is the maximum number of people that could possibly fit a single cell. Using (10.6) in (10.5) we get modified one-equation model given by

$$\frac{\partial \rho(x,t)}{\partial t} + \sum_{i=1}^{2} \frac{\partial}{\partial x_i}\left[v_{if}(x,t)\left(1 - \frac{\rho(x,t)}{\rho_{\max}}\right)\rho(x,t)\right] = 0 \tag{10.7}$$

subject to conditions (10.2) and (10.3).

10.2.2 Control Model

To formulate the control problem we need to choose a control variable. To do so we use Greenshields model to represent the relationship between traffic density and the velocity field given by (10.6). In this model we take free flow velocity vector fields $v_{if}(x,t)$ as the distributed control variables denoted by $u_i(x,t)$. This gives us following representation of the control system

$$\frac{\partial \rho(x,t)}{\partial t} = -\sum_{i=1}^{2} \frac{\partial}{\partial x_i}\left[u_i(x,t)\left(1 - \frac{\rho(x,t)}{\rho_{\max}}\right)\rho(x,t)\right] \tag{10.8}$$

where $u_i(x,t) \in H[\Omega, \Re]$ are the control variables.

10.2 Feedback Control of One-Equation Model

10.2.3 State Feedback Control

Here we address the problem of synthesizing a distributed state feedback controller $u(x,t)$ that stabilizes origin $\rho(x,t) = 0$ of system (10.8). We use the method of feedback linearization for PDEs as in Chap. 9. More specifically we consider control law of the form

$$u_i(x,t) = F_i(\rho(x,t))$$

which makes origin of the closed loop dynamics exponentially stable. Here $F_i(\rho)$ is a nonlinear operator mapping $H^2[\Omega, \Re]$ into $H^1[\Omega, \Re]$. Designing the operator $F_i(\rho)$ as

$$F_i(\rho(x,t)) = -\left[\left(1 - \frac{\rho(x,t)}{\rho_{\max}}\right)\rho(x,t)\right]^{-1} k\frac{\partial \rho(x,t)}{\partial x_i} \qquad (10.9)$$

we get the following closed loop dynamics

$$\frac{\partial \rho(x,t)}{\partial t} - k\left[\frac{\partial^2}{\partial x_1^2} + \frac{\partial^2}{\partial x_2^2}\right]\rho(x,t) = 0$$

or

$$\frac{\partial \rho(x,t)}{\partial t} - k\nabla^2 \rho(x,t) = 0 \qquad (10.10)$$

with boundary conditions given by (10.2) and (10.3).

10.2.4 Lyapunov Stability Analysis

The stability problem is to establish sufficient conditions for which the origin of the closed loop dynamics (10.8) is exponentially stable. In order to check the stability of the closed loop system we use Lyapunov function analysis. For the 2-D system, definition of stability in terms of Lyapunov can be established in a similar way to the 1-D case as given in Sect. 9.3.3.

The determination of conditions for which estimate (9.17) is satisfied amounts to establishing conditions for which the null state of the closed loop system (10.10) is exponentially stable with respect to the L_2 norm given by

$$\|\rho\|_2 \to \left[\int_\Omega |\rho|^2 \, d\Omega\right]^{1/2} \qquad (10.11)$$

where exponential stability implies $\|\rho(t)\|_2 \to 0$ as $t \to \infty$. Let us consider the same Lyapunov functional $V(t)$ as given by (9.21) for the system (10.10). Here $V : H^2[\Omega] \to \Re_+$ is a smooth functional of the form

$$V(t) = \tfrac{1}{2}\|\rho(t,x)\|_2^2 = \tfrac{1}{2}\int_\Omega |\rho(t,x)|^2 d\Omega \tag{10.12}$$

Differentiating with respect to time, we get the time rate of change of as

$$\frac{dV(t)}{dt} = -D\int_\Omega \sum_{i=1}^{2}\left(\frac{\partial \rho(x,t)}{\partial x_i}\right)^2 d\Omega$$

Now we make use of Sobolev Inequality (9.21) which for the 2-D case is

$$\int_\Omega \sum_{i=1}^{2}\left(\frac{\partial \rho}{\partial x_i}\right)^2 d\Omega \geq C^{-1}\int_\Omega \rho^2 d\Omega$$

where C is a positive real number. The rate of change of $V(t)$ can be thus be bounded by

$$\frac{dV(t)}{dt} \leq -DC^{-1}\int_\Omega \rho^2 d\Omega = -2DC^{-1}V(t) = -\beta V(t)$$

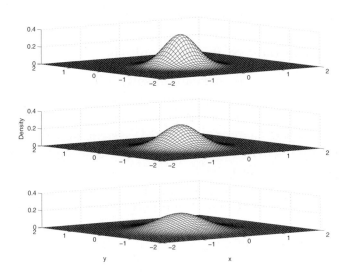

Fig. 10.1. Density response at different time instants for one-equation model

10.2 Feedback Control of One-Equation Model

It follows that $V(t) \leq V(t_0)e^{-\beta(t-t_0)}$ or

$$\|\rho(t,x)\|_2 \leq \|\rho(t_0,x)\|_2 \, e^{-\beta(t-t_0)}$$

with $\beta = 2kC^{-2}$. As long as $\beta > 0$, null state of (10.10) is exponentially stable and condition (9.17), $\|U(t)\|_2 \leq e^{-b(t-t_0)}$ is satisfied with $b = \beta$. Thus equilibrium of closed loop system (10.10) using feedback control (10.9) is exponentially stable.

10.2.5 Simulation Results

This section shows simulation results for the closed loop system (10.10) using control law (10.9 and 10.10). The simulation results are shown in Fig. 10.1 and Fig. 10.2. The numerical method used is the Lax-Friedrichs scheme [67]. For simulation the initial distribution of density is considered to be Gaussian. The initial condition is given by

$$\rho(x,0) = G\exp(-(x_1-a)^2 - (x_2-b)^2)$$

with (a,b) being the centre of the Gaussian distribution and G, the highest magnitude of the distribution. The control action response

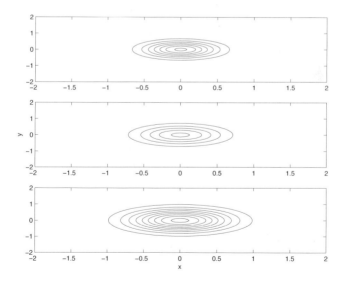

Fig. 10.2. Contours of the density response at different time instants for one-equation model

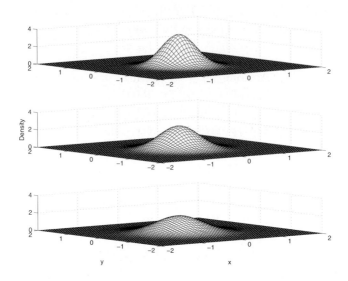

Fig. 10.3. Density response at different time instants for two-equation model

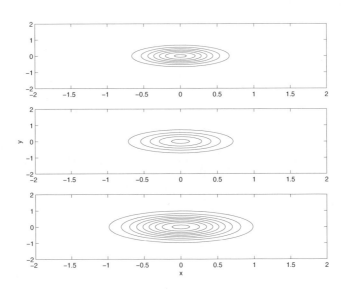

Fig. 10.4. Contours of the density response at different time instants for two-equation model

for a corridor is shown in Fig. 10.3. The simulation shows the density response as mesh plot snapshots. After some finite time the density has decreased to zero. The second simulation is shown as contour plot snapshots in Fig. 10.4. As seen from plot, density at every point in space is decreasing exponentially with time.

10.3 Feedback Control for Two-Equation Model

In this section we design a feed back control for the two-equation model of the evacuation system using backstepping approach. The Lyapunov functional (10.15) which was used as a stability analysis tool for one-equation model will be used as a feedback control design tool for this system. The design of feed back control is done in such a way that Lyapunov functional or its derivative has certain properties that guarantee boundedness or convergence to an equilibrium point.

10.3.1 Two Equation Model

In this section we consider a higher order model or more precisely a system of two partial differential equations for a 2-D corridor. This model consists of conservation of mass equation coupled with a second set of equations based on the principle of conservation of momentum. The first equation is the conservation of mass equation (10.5). The second set of equations is derived from conservation of momentum for 2-D flow [3] given by

$$\frac{\partial \left(\rho\left(x,t\right) v_i\left(x,t\right) \right)}{\partial t} + \mathrm{div}\left(\rho v_i v \right) = -\frac{\partial p\left(x,t\right)}{\partial x_i}, \quad i = 1, 2$$

The x_1 and x_2 components of momentum equation can be written as

$$\frac{\partial \left(\rho\left(x,t\right) v_1\left(x,t\right) \right)}{\partial t} + \frac{\partial \left(\rho\left(x,t\right) v_1\left(x,t\right)^2 \right)}{\partial x_1}$$
$$+ \frac{\partial \left(\rho\left(x,t\right) v_1\left(x,t\right) v_2\left(x,t\right) \right)}{\partial x_2} = -\frac{\partial p\left(x,t\right)}{\partial x_1}$$

and

$$\frac{\partial (\rho(x,t) v_1(x,t))}{\partial t} + \frac{\partial \left(\rho(x,t) v_2(x,t)^2\right)}{\partial x_2} + \frac{\partial (\rho(x,t) v_1(x,t) v_2(x,t))}{\partial x_1} = -\frac{\partial p(x,t)}{\partial x_2}$$

where $p(x,t) \in H^1[(0,L), \Re]$ is pressure. The relationship between density and flow is given by $q_i(x,t) = \rho(x,t) v_i(x,t)$. Thus we have the dynamics of an evacuation system given by following two-equation model

$$\frac{\partial \rho(x,t)}{\partial t} + \sum_{i=1}^{2} \frac{\partial q_i(x,t)}{\partial x_i} = 0 \qquad (10.13)$$

and

$$\frac{\partial q_1(x,t)}{\partial t} + \frac{\partial}{\partial x_1}\left(\frac{q_1(x,t)^2}{\rho(x,t)}\right) + \frac{\partial}{\partial x_2}\left(\frac{q_1 q_2(x,t)}{\rho(x,t)}\right) = -\frac{\partial p}{\partial x_1}$$

$$\frac{\partial q_2(x,t)}{\partial t} + \frac{\partial}{\partial x_2}\left(\frac{q_2(x,t)^2}{\rho(x,t)}\right) + \frac{\partial}{\partial x_1}\left(\frac{q_1 q_2(x,t)}{\rho(x,t)}\right) = -\frac{\partial p}{\partial x_2} \quad (10.14)$$

with initial conditions

$$\rho(x,t_0) = \rho_0(x), q(x,t_0) = q_0(x) \qquad (10.15)$$

and subject to boundary conditions

$$\rho(x,t) = 0 \; \forall \, t \in [0,\infty), x \in \partial\Omega$$

$$q(x,t) = 0 \; \forall \, t \in [0,\infty), x \in \partial\Omega \qquad (10.16)$$

10.4 Control Model

In the two equation model (10.13) and (10.14) we choose divergence of pressure $\frac{\partial p(x,t)}{\partial x_i}$, as distributed control variables $u_i(x,t)$ which give us the following control model

$$\frac{\partial \rho(x,t)}{\partial t} = -\sum_{i=1}^{2} \frac{\partial q_i(x,t)}{\partial x_i}$$

10.4 Control Model

and

$$\frac{\partial q_1(x,t)}{\partial t} = -\frac{\partial}{\partial x_1}\left(\frac{q_1(x,t)^2}{\rho(x,t)}\right) - \frac{\partial}{\partial x_2}\left(\frac{q_1 q_2(x,t)}{\rho(x,t)}\right) + u_1(x,t)$$

$$\frac{\partial q_2(x,t)}{\partial t} = -\frac{\partial}{\partial x_2}\left(\frac{q_2(x,t)^2}{\rho(x,t)}\right) - \frac{\partial}{\partial x_1}\left(\frac{q_1 q_2(x,t)}{\rho(x,t)}\right) + u_2(x,t)$$

This system can be rewritten in the form

$$\frac{\partial \rho(x,t)}{\partial t} = -\sum_{i=1}^{2} \frac{\partial q_i(x,t)}{\partial x_i} \qquad (10.17)$$

$$\frac{\partial q_i(x,t)}{\partial t} = \bar{u}_i(x,t); i = 1, 2 \qquad (10.18)$$

where

$$\bar{u}_1(x,t) = -\frac{\partial}{\partial x_1}\left(\frac{q_1(x,t)^2}{\rho(x,t)}\right) - \frac{\partial}{\partial x_2}\left(\frac{q_1 q_2(x,t)}{\rho(x,t)}\right) + u_1(x,t)$$

and

$$\bar{u}_2(x,t) = -\frac{\partial}{\partial x_2}\left(\frac{q_2(x,t)^2}{\rho(x,t)}\right) - \frac{\partial}{\partial x_1}\left(\frac{q_1 q_2(x,t)}{\rho(x,t)}\right) + u_2(x,t)$$

are the new control variables.

10.4.1 State Feedback Control Using Backstepping

Here we address the problem of synthesizing a distributed state feedback controller $u(x,t) = [u_1(x,t), u_2(x,t)]$ that stabilizes origin $\rho(x,t) = 0, q(x,t) = 0$ of control system (10.17 and 10.18). More specifically we consider control law

$$\bar{u}_i(x,t) = \bar{F}_i(\rho(x,t), q(x,t))$$

such that origin of closed loop dynamics is exponentially stable. $\bar{F}_i$ is a nonlinear operator mapping $H^2[\Omega, \Re]$ into $H^1[\Omega, \Re]$. The control strategy adopted here is similar to feedback control by backstepping PDEs for 1-D case which is extended to the 2-D case here. The sequence of steps in control design is the same as before and is given as follows:

1. First we design control law for (10.17) where $q(x,t)$ can be viewed as an input. We proceed to design a conceptual control law $q_i(x,t) = \bar{G}_i(\rho(x,t))$ to stabilize origin $\rho(x,t) = 0$ as

$$q_i(x,t) = \bar{G}_i(\rho(x,t)) = \int_0^{x_i} \frac{\partial^2 \rho(x,t)}{\partial m^2} dm, \quad i = 1, 2 \quad (10.19)$$

$\bar{G}_i$ is a nonlinear operator mapping $H^2[\Omega, \Re]$ into $H^2[\Omega, \Re]$. The conceptual closed loop dynamics for (10.17) are

$$\frac{\partial \rho(x,t)}{\partial t} = D\left[\frac{\partial^2}{\partial x_1^2} + D\frac{\partial^2}{\partial x_2^2}\right]\rho(x,t)$$

or

$$\frac{\partial \rho(x,t)}{\partial t} = D\nabla^2 \rho(x,t) \quad (10.20)$$

which is similar to (10.10). As we have already shown the origin of this equation is asymptotically exponentially stable. In addition there exists a conceptual Lyapunov functional

$$V(t) = \tfrac{1}{2}\|\rho(t,x)\|_2^2 = \tfrac{1}{2}\int_\Omega |\rho(t,x)|^2 d\Omega$$

for this system which satisfies $\frac{dV(t)}{dt} \leq -\beta V(t)$ and ensures its stability.

2. From the knowledge of this conceptual Lyapunov function $V(t)$ we design a smooth feedback control to stabilize the origin of the overall system. We therefore add to the conceptual Lyapunov function (10.12) a term penalizing the difference between q and $G(\rho)$. For this purpose we rewrite the dynamics (10.17) as

$$\frac{\partial \rho(x,t)}{\partial t} = -\sum_{i=1}^{2} \frac{\partial \bar{G}_i(\rho(x,t))}{\partial x_i} - \sum_{i=1}^{2} \frac{\partial}{\partial x_i}(q_i(x,t) - \bar{G}_i(\rho(x,t)))$$

Defining error variables $z_i(x,t) = q_i(x,t) - \bar{G}_i(\rho(x,t))$ result in the following modified dynamics

$$\frac{\partial \rho(x,t)}{\partial t} = -\sum_{i=1}^{2} \frac{\partial \bar{G}_i(\rho(x,t))}{\partial x_i} - \sum_{i=1}^{2} \frac{\partial z_i(x,t)}{\partial x_i} \quad (10.21)$$

$$\frac{\partial z_i(x,t)}{\partial t} = u_{ni}(x,t) \quad (10.22)$$

10.4 Control Model

where $u_{ni}(x,t) = \bar{u}_i(x,t) - \frac{\partial G_i(\rho(x,t))}{\partial t}$, $i = 1, 2$ are the new control variables and $z_i(x,t) \in H^2[(\Omega, \Re]$.

3. Now let us modify the Lyapunov functional (10.12) for the overall system as

$$V_a(t) = V(t) + \frac{1}{2}\|z(x,t)\|_2^2 = \frac{1}{2}\int_0^L |\rho(x,t)|^2 \, dx \qquad (10.23)$$
$$+ \frac{1}{2}\int_\Omega \sum_{i=1}^2 |z_i(x,t)|^2 \, d\Omega$$

where $z(x,t) = [z_1(x,t), z_2(x,t)]$. The time rate of change of this functional using (10.21) and (10.22) is given as

$$\frac{dV_a(t)}{dt} = \int_\Omega \rho(x,t) \nabla^2 \rho(x,t) \, d\Omega + \int_\Omega \sum_{i=1}^2 \rho(x,t) \frac{\partial z_i(x,t)}{\partial x_i} \, d\Omega$$
$$+ \int_\Omega \sum_{i=1}^2 z_i(x,t) u_{ni}(x,t) \, d\Omega$$

we know that the first term is bounded by $-\beta V(t)$. Therefore

$$\frac{dV_a(t)}{dt} \leq -\beta V(t) + \int_\Omega \sum_{i=1}^2 \rho(x,t) \frac{\partial z_i(x,t)}{\partial x_i} \, d\Omega$$
$$+ \int_\Omega \sum_{i=1}^2 z_i(x,t) u_{ni}(x,t) \, d\Omega$$

We have to choose a new control law $u_{ni}(x,t)$ in such a manner that the time derivative of new functional or the sum of second and third terms is also bounded by some negative definite function. By choosing the control law

$$u_{ni}(x,t) = -\hat{k} z_i(x,t) - z_i(x,t)^{-1} \rho(x,t) \frac{\partial z_i(x,t)}{\partial x_i} \qquad (10.24)$$

we get the following result

$$\frac{dV_a(t)}{dt} \leq -\beta V(t) - \hat{k}\|z(x,t)\|_2^2 = -2\beta V_a(t)$$

with $\hat{k}=2\beta>0$. This shows that the origin $\rho(x,t)=0$, $z(x,t)=0$ of closed loop system (10.21) and (10.22) is exponential stable.

The feedback control laws are given by the following partial differential-integral equations

$$u_1(x,t) = u_{n1}(x,t) + \frac{\partial G_1(\rho(x,t))}{\partial t} + \frac{\partial}{\partial x_1}\left(\frac{q_1(x,t)^2}{\rho(x,t)}\right) + \frac{\partial}{\partial x_2}\left(\frac{q_1 q_2(x,t)}{\rho(x,t)}\right) \quad (10.25)$$

and

$$u_2(x,t) = u_{n2}(x,t) + \frac{\partial G_2(\rho(x,t))}{\partial t} + \frac{\partial}{\partial x_2}\left(\frac{q_1(x,t)^2}{\rho(x,t)}\right) + \frac{\partial}{\partial x_1}\left(\frac{q_1 q_2(x,t)}{\rho(x,t)}\right) \quad (10.26)$$

with $u_{ni}(x,t)$ given by (10.24) as

$$u_{ni}(x,t) = -\hat{k}z_i(x,t) - z_i(x,t)^{-1}\rho(x,t)\frac{\partial z_i(x,t)}{\partial x_i}$$

where $z_i(x,t) = q_i(x,t) - G_i(\rho(x,t))$. Hence the closed loop dynamics for two equation model are exponentially stable with this control law.

10.4.2 Simulation Results

Here we show simulation results for the closed loop system (10.17) and (10.18) using controller (10.25) designed in previous section. The simulation results showing the density response is plotted in Fig. 10.3 and Fig. 10.4 and the flow response is plotted in Fig. 10.5 and Fig. 10.6. The numerical technique used to simulate the system is same as before. For simulation the initial distribution for both density and flow is considered to be Gaussian. As seen from the plots after some finite time, the density and flows in both directions have decreased to zero.

10.4 Control Model

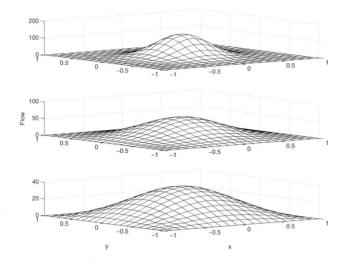

Fig. 10.5. Flow response at different time instants for two-equation model

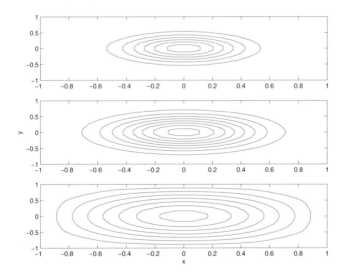

Fig. 10.6. Contours of the flow response at different time instants for two-equation model

10.5 Exercises

Problem 10.5.1 Change the components of control (10.9) to the following control law

$$F_i(\rho(x,t)) = \frac{1}{\rho(x,t)} \frac{\partial \rho(x,t)}{\partial x_i}$$

Apply this control to system (10.8). Find the closed loop dynamics and repeat the stability analysis.

Problem 10.5.2 Add the term $1/\rho_{\max} \partial \rho(x,t)/\partial x$ to each component of control in above problem and repeat the Lyapunov Analysis. Is the system still exponentially stable?

Problem 10.5.3 Work out the details of Sobolev inequality for (10.20).

10.6 Computer Code

In this section the code used for the above simulations is included. The code is in matlab and consists of m-file named feedback_2d.

10.6.1 feedback_2d

```
n=20;%points in x direction
l=20;%points in y direction
m=600;%time samples
T=3;
L=2;
hx=L/(n-1);  % spacing in x-direction
hy=L/(l-1);
ht=T/(m-1);  % spacing in time
s=ht/(hx^2);
r=ht/(hy^2);
x=[-1:hx:1];
y=[-1:hy:1];
t=[0:ht:3]
rm=1000;
```

10.6 Computer Code

```
x_max=1;
y_max=1;
[x,y] = meshgrid(-x_max:hx:x_max, -y_max:hy:y_max);
R=sqrt(x.^2+y.^2);
std_R=x_max/4;
mean_R=0;
Row_0 = 50*exp(-(R-mean_R).^2./.........
(2*std_R^2))/ (std_R*sqrt(2*pi));
B=0;
Row(1,1:n,1:l)=Row_0;
for i=1:m-1
    for j=2:n-1
      for k=2:l-1
    Row(i+1,j,k)=Row(i,j,k)+s*(Row(i,j+1,k)-2*Row(i,j,k)
    +Row(i,j-1,k)).....
    +r*(Row(i,j,k+1)-2*Row(i,j,k)+Row(i,j,k-1));
        end

    end
end

for i=1:m
    for j=1:n
      for k=1:l
            if i == 1
          A(j,k)=Row(i+1,j,k)
         end
          if i == 5
          B(j,k)=Row(i+1,j,k)
          end
          if i == 10
         C(j,k)=Row(i+1,j,k)
          end
          if i == 20
         D(j,k)=Row(i+1,j,k)
         end
          if i == 25
         E(j,k)=Row(i+1,j,k)
```

```
            end
        end

      end
end

figure(1)
subplot(4,1,1),
mesh(x,y,A)

subplot(4,1,2),
mesh(x,y,B)
subplot(4,1,3),
mesh(x,y,C)

subplot(4,1,4),
mesh(x,y,D)


figure(2)
subplot(4,1,1),
[c,h1] = contour(x,y,A); clabel(c,h1,'manual');
subplot(4,1,2),
[c,h1] = contour(x,y,B); clabel(c,h1,'manual');
subplot(4,1,3),
[c,h1] = contour(x,y,C); clabel(c,h1,'manual');
subplot(4,1,4),
[c,h1] = contour(x,y,D); clabel(c,h1,'manual');

a=1./(1-2*row/row_m);
b=(del_row);
c=(del2_row);

vf=(a/b)*c;
v=vf.*(1-row/row_m);

for i=1:m
 for j=1:n
```

10.6 Computer Code

```
      if v(i,j)<0
      v(i,j)=0;
      end
  end
end
q1=row.*v;
```

Chapter 11

Robust Feedback Control

This chapter presents design of robust nonlinear feedback controllers for two different models representing evacuation dynamics in 2-D. The models presented here are similar to the ones discussed in Chap. 9. We address the feedback control problem for both models.

11.1 Introduction

In previous chapters we addressed the control of crowd dynamic systems without accounting for the presence of uncertainty in the design of the controller. The uncertainty is a mismatch between the model used for controller design and the actual process model. The uncertain function may include uncertain model parameters or external disturbances. Here we consider the case where we have an uncertainty in the input to the system. The uncertainty is due to the mismatch between the control command and the actual control command followed by people and is distributed in space. The objective is to develop a framework for the synthesis of distributed robust controllers that handle the effect of this uncertain variable. A distributed robust controller is derived that guarantees boundedness of state and achieves asymptotic stabilization with arbitrary degree of asymptotic attenuation of the effect of uncertain variables on the output of the closed-loop system. The controller is designed constructively using Lyapunov's direct method [59] and requires the existence of known bounding functions that capture the magnitude of the uncertain term. This chapter presents the design of robust nonlinear feedback controllers for two models given in Chap. 9 representing crowd dynamics in 1-D with uncertainty in the control input. The models are based on the laws of conservation of mass and momentum and are given in Sect. 9.2.1. In both cases the objective of control design is to synthesize a nonlinear distributed feedback controller that stabilizes the system and guarantees stability of the closed loop system in presence of uncertainty. The organization of this chapter is as follows. In Sect. 11.2 we formulate the uncertain control model and present robust control design for the one-equation crowd model. This section also studies Lyapunov stability for this model. Simulation results for closed loop dynamics are presented where the developed control method is tested with a disturbance in the system. Section 11.3 presents the robust control and stability analysis for the two-equation system model.

11.2 Feedback Control for Continuity Equation Model

In this section we formulate the uncertain control model and present robust control design for the one-equation model. The problem is to design a state feedback control that guarantees the desired performance of the closed-loop system irrespective of uncertain elements. The one-equation model is given in Sect. 9.2.1 by (9.7) as

$$\frac{\partial(\rho(x,t))}{\partial t} + \frac{\partial(q(x,t))}{\partial x} = 0 \qquad (11.1)$$

Here $\rho(x,t)$ is the variable we want to control. The flow $q(x,t)$ is obtained as a product of density and velocity as $q(x,t) = \rho(x,t)v(x,t)$ and the velocity–density relationship as given by Greenshield's model is

$$v = v_f(1 - \rho/\rho_{\max}) \qquad (11.2)$$

where $v_f = v_f(x,t)$ is the free flow speed and $\rho_{\max}$ is the maximum or jam density. The one-equation model is therefore given by

$$\frac{\partial \rho}{\partial t} + \frac{\partial}{\partial x}(\rho v_f(1 - \rho/\rho_{\max})) = 0 \qquad (11.3)$$

11.2 Feedback Control for Continuity Equation Model

By choosing free flow velocity vector field $v_f = v_f(x,t)$ as the distributed control variable denoted by u we get the following control model

$$\frac{\partial \rho}{\partial t} + \frac{\partial}{\partial x}(\rho u(1 - \rho/\rho_{\max})) = 0 \qquad (11.4)$$

11.2.1 Input Uncertain Control Model

Let us consider that the uncertain variable for this system is the control input or the free flow velocity. This means that there is an error in the control command and the actual input command followed by people. The uncertain model is therefore given as

$$\frac{\partial \rho}{\partial t} + \frac{\partial}{\partial x}[(1 - \rho/\rho_{\max})\rho(u + \theta)] = 0 \qquad (11.5)$$

In above equation denotes the unknown function which takes care of uncertainty in the input to the system. The uncertain term here satisfies an important structural property namely it enters the state equation exactly at the point where the control variable enters. This property will be referred to as the matching condition. The nominal model (the system without uncertainty) for this system is described by

$$\frac{\partial \rho}{\partial t} + \frac{\partial}{\partial x}[(1 - \rho/\rho_{\max})\rho u] = 0$$

The first step is to design a stabilizing feedback controller for the nominal model. With the feedback controller (9.8) $u = F(\rho)$ given in Chap. 9 as

$$F(\rho(x,t)) = -[(1 - \rho/\rho_{\max})\rho]^{-1} D \frac{\partial \rho}{\partial x} \qquad (11.6)$$

we get the nominal closed-loop system as

$$\frac{\partial \rho}{\partial t} - D \frac{\partial^2 \rho}{\partial x^2} = 0 \qquad (11.7)$$

Thus with $|\theta(t, u, \rho)| = 0$ and $u = F(\rho)$ the nominal closed loop model (11.7) has exponentially stable origin and there exists a Lyapunov function

$$V(t) = \tfrac{1}{2} \|\rho\|_2^2 = \tfrac{1}{2} \int_0^L |\rho|^2 dx$$

which satisfies $\frac{dV(t)}{dt} \leq -\beta V(t) = -2DC^{-1} \|\rho\|_2^2$ with $\beta = 2D$.

11.2.2 Robust Control by Lyapunov Redesign Method

In this section we consider the system of (11.5) and address the problem of synthesizing a distributed state feedback controller that stabilizes the closed-loop system irrespective of the uncertainty. The controller is designed constructively using Lyapunov's direct method where we use the Lyapunov function for the system to design feedback control. A standard method for finding a Lyapunov function for an uncertain system is developed in [19] and is known as Lyapunov redesign. This technique has been incorporated in various books like [27, 59]. The key idea of this method is to employ a Lyapunov function for the nominal system as Lyapunov function for the uncertain system. This re-use of the same Lyapunov function is referred to by the term "redesign". The Lyapunov redesign technique uses a Lyapunov functional of a nominal system to design an additional control component to robustify the design to a class of large uncertainties that satisfy the matching condition; i.e., the uncertain terms enter the state equation at the same point as the input. Lyapunov redesign can be used to achieve robust stabilization. The goal is to design feedback control law for (11.5) as

$$u = F(\rho) + G(\rho) \tag{11.8}$$

such that we achieve closed loop stability and asymptotic attenuation of $\theta(t, u, \rho)$ where $G(\rho)$ is a nonlinear operator mapping $H^2[(0, L), \Re]$ into $H^1[(0, L), \Re]$. In (11.8), $F(\rho)$ achieves closed loop stability and $G(\rho)$ asymptotically attenuates the effect of $\theta(t, u, \rho)$. The function $F(\rho)$ will be designed by the previous approach and $G(\rho)$ will be designed using Lyapunov redesign method. The design of function $G(\rho)$ is known as Lyapunov redesign [59] and is done on the basis of the assumption that we have a bounding function that captures the size of the disturbance. Let us assume with controller (11.8) there exists a known smooth function which bounds the magnitude of uncertain variables as:

$$\|\theta(t, \rho, u)\| = \|\theta(t, \rho, (F(\rho) + G(\rho)))\| \leq \gamma(t, \rho) + \kappa \|G(\rho)\| \tag{11.9}$$

where γ is a nonnegative H^1 function and is a measure of size of uncertainty $\theta(t, u, \rho)$. From estimate (11.9), the only requirement is

11.2 Feedback Control for Continuity Equation Model

the knowledge of γ which doesn't necessarily have to be small. From the knowledge of Lyapunov function $V(t)$ and functions γ and κ the goal is to design $G(\rho)$ and apply $u = F(\rho) + G(\rho)$ to the actual system (11.5) such that the overall closed loop system is stabilized in presence of uncertainty. By using the control law (11.6) the feedback control law (11.8) is given by

$$u = -[(1 - \rho/\rho_{\max})\rho]^{-1} D\frac{\partial \rho}{\partial x} + G(\rho) \quad (11.10)$$

Under the feedback control law (11.10) the closed-loop dynamics for (11.5) take the form

$$\frac{\partial \rho}{\partial t} - D\frac{\partial^2 \rho}{\partial x^2} - \frac{\partial}{\partial x}[\rho(1 - \rho/\rho_{\max})G(\rho)] = 0 \quad (11.11)$$

Thus the error in the closed loop dynamics (11.10) and (11.11) due to input uncertainty is given by

$$\theta = \frac{\partial}{\partial x}[\rho(1 - \rho/\rho_{\max})G(\rho)] \quad (11.12)$$

Let us denote the component $G(\rho)$ of control input (11.8), by v and $w(\rho) = \rho(1 - \rho/\rho_{\max})$ we can rewrite (11.12) as

$$\theta = -\frac{\partial(wv)}{\partial x} = v(\partial w/\partial x) - w(\partial v/\partial x) \quad (11.13)$$

where $\partial w/\partial x = (1 - 2\rho/\rho_{\max})\partial \rho/\partial x$. The magnitude of this uncertain term with respect to L_2 norm is given by

$$\|\theta(t, \rho, u)\|_2^2 = \int_0^L |v(\partial w/\partial x)|^2 \, dx + \int_0^L |w(\partial v/\partial x)|^2 \, dx$$

$$+ 2\int_0^L |v(\partial w/\partial x)| \, |w(\partial v/\partial x)| \, dx \quad (11.14)$$

The bound on the magnitude of uncertainty (11.14) can be found by applying Holders Inequality [25] to each integral term in the above equation, which results in

$$\|\theta(t, \rho, u)\|_2^2 \leq \|v\|_2^2 \, \|(\partial w/\partial x)\|_2^2 + \|w\|_2^2 \, \left\|\left(\frac{\partial v}{\partial x}\right)\right\|_2^2$$

$$+ 2\left\|v\left(\frac{\partial w}{\partial x}\right)\right\|_2 \left\|w\left(\frac{\partial v}{\partial x}\right)\right\|_2$$

Therefore from the above inequality we have the following bound on uncertainty

$$\|\theta(t,\rho,u)\|_2 \le \|v\|_2 \left\|(\frac{\partial w}{\partial x})\right\|_2 + \|w\|_2 \left\|(\frac{\partial v}{\partial x})\right\|_2$$

$$= \gamma(\rho,t) + \kappa(\rho,t)\bar{v} \qquad (11.15)$$

where $\kappa(\rho,t) = \|(\partial w/\partial x)\|_2$, $\gamma(\rho,t) = \|w\|_2$ and $\bar{v} = \|v\|_2/\|(\partial v/\partial x)\|_2$. This bound will be utilized to design the control law. We have the closed-loop dynamics

$$\frac{\partial \rho}{\partial t} + \frac{\partial}{\partial x}[\rho(1-\rho/\rho_m)(F(\rho))] + \frac{\partial}{\partial x}[\rho(1-\frac{\rho}{\rho_m})(G(\rho)+\theta)] = 0 \quad (11.16)$$

As can be seen the system (11.16) is a perturbation of the nominal closed-loop system (11.7), the third term being the perturbation. Let us choose the Lyapunov function $V(t)$ for (11.7) same as before. To design $G(\rho)$ we find the rate of change of Lyapunov function as

$$\frac{dV(t)}{dt} = \int_0^L \rho \frac{\partial \rho}{\partial t} dx$$

$$\le -k\|\rho\|_2^2 + \int_0^L \rho \frac{\partial}{\partial x}[\rho(1-\rho/\rho_m)(G(\rho)+\theta)]dx \qquad (11.17)$$

The first inequality is because of the asymptotic stability of nominal closed-loop system (11.7). We need to choose a control law $G(\rho)$ so as to cancel the destabilizing effort of $\theta(t,u,\rho)$ on $dV(t)/dt$. The law $G(\rho)$ should be such that the second term in (11.17) is negative semi-definite in order to have asymptotic stability. By using $w(\rho) = \rho(1-\rho/\rho_m)$ and denoting $G(\rho)$ by v we can rewrite (11.17) as

$$\frac{dV(t)}{dt} \le -k\|\rho\|_2^2 + \int_0^L \rho \frac{\partial(wv)}{\partial x}dx + \int_0^L \rho \frac{\partial(w\theta)}{\partial x}dx \qquad (11.18)$$

Since the disturbance in our case is a function of space and time that is $\theta = \theta(t,u,\rho)$, let us define $\hat{\theta} = \partial(w\theta)/\partial x$ and a new control

11.2 Feedback Control for Continuity Equation Model

variable $\hat{v} = \partial(wv)/\partial x$. Therefore we have

$$\frac{dV(t)}{dt} \leq -k \|\rho\|_2^2 + (\int_0^L \rho \hat{v} dx + \int_0^L \rho \hat{\theta} dx) \tag{11.19}$$

Now let us apply the Holders inequality to the second integral in the second term of (11.19). This results in the following

$$\frac{dV(t)}{dt} \leq -k \|\rho\|_2^2 - (\int_0^L \rho \hat{v} dx + \|\rho\|_2 \|\hat{\theta}\|_2) \tag{11.20}$$

By using Sobolev inequality (9.16) we can show that $\|\hat{\theta}\|_2 = \|\partial(w\theta)/\partial x\|_2 \geq C^{-1} \|w\theta\|_2$ and by using (11.15) we have the rate of change of Lyapunov as

$$\frac{dV(t)}{dt} \leq -k \|\rho\|_2^2 - \left(\int_0^L \rho \hat{v} dx + \|\rho\|_2 \|w\|_2 (\gamma + \kappa \bar{v})\right) \tag{11.21}$$

Now let us choose the state feedback control as following

$$\hat{v} = \frac{-\eta(\rho, t)}{\kappa(\rho, t)} w(x, t)$$

or

$$v = \frac{1}{w} \int_0^L \left(\frac{-\eta(\rho, t)}{\kappa(\rho, t)} w(x, t)\right) dx \tag{11.22}$$

where $\eta(\rho, t) \geq \gamma(\rho, t)$ is a nonnegative function. By using this control law we can show that $\bar{v} = \|v\|_2 / \|\partial v / \partial x\|_2 \geq \|v\|_2 / C \|v\|_2 = C^{-1}$. Using this relation we can write (11.21) as

$$\frac{dV(t)}{dt} \leq -k \|\rho\|_2^2 - \left(\int_0^L \rho \hat{v} dx + C^{-1} \|\rho\|_2 \|w\|_2 (\gamma + \kappa C^{-1})\right) \tag{11.23}$$

By using the control law (11.22) and then applying Holders inequality to the integral term in above equation it becomes $\int_0^L \rho \hat{v} dx \leq C \|\rho\|_2 \|w\|_2 \|\eta/k\|_2$. Hence (11.23) can be written as

$$\frac{dV(t)}{dt} \leq -k \|\rho\|_2^2 - C^{-1} \|\rho\|_2 \|w\|_2 \gamma(\rho, t)$$

$$+ C^{-1} \|\rho\|_2 \|w\|_2 \left(C \left\| \frac{\eta}{k} \right\|_2 - \kappa C^{-1} \right)$$

As long as $(C \|\eta/k\|_2 - \kappa C^{-1}) \leq \gamma$, we have asymptotic stability in presence of the disturbance. The robust control law for (11.5) is therefore given as

$$u = -[(1 - \rho/\rho_m)\rho]^{-1} D \frac{\partial \rho}{\partial x} + \frac{1}{w} \int_0^L \left(\frac{-\eta(\rho, t)}{\kappa(\rho, t)} w(x, t) \right) dx \quad (11.24)$$

11.2.3 Simulation Results

This section shows simulation results for the system (11.5) using the robust control law (11.26). For simulation the initial distribution of density is considered to be Guassian. The simulation results are shown in Figs. 11.1, 11.2 and 11.3. The density response for nominal control model (model without uncertainty) is shown in Fig. 11.1. In Fig. 11.2 the density response is shown as a contour plot which shows people moving towards the right (exit) of the corridor. The simulation in Fig. 11.1 shows the density response at different time instants where the response flattens with time. The density response for uncertain model is shown in Fig. 11.2. We have added a step function (with respect to time over all distance) disturbance to the system. As seen from the plot in Fig. 11.2. the density is moving back because of the disturbance. In Fig. 11.3. we show the response with the robust controller (11.26) added to the system. As is seen from the figure the response has improved and the effect of disturbance is cancelled.

11.2 Feedback Control for Continuity Equation Model

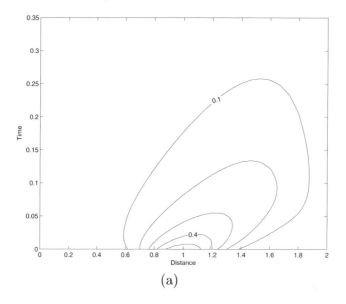

(a)

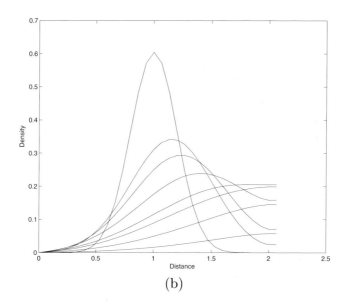

(b)

Fig. 11.1. Density response for one-equation nominal-model: (a) Contours of the response; (b) Density at different time instants

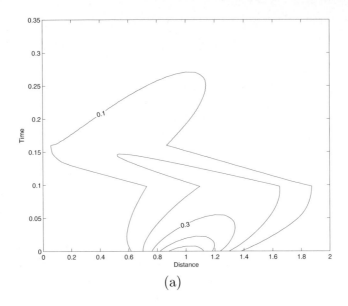

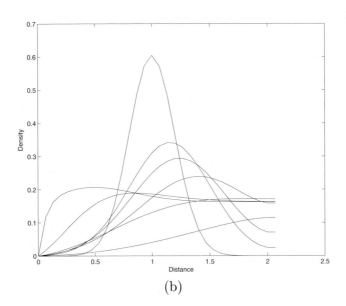

Fig. 11.2. Density response for one-equation uncertain-model: **(a)** Contours of the response; **(b)** Density at different time instants

11.2 Feedback Control for Continuity Equation Model

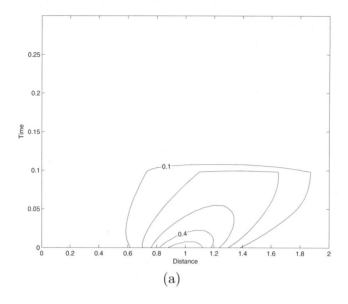

(a)

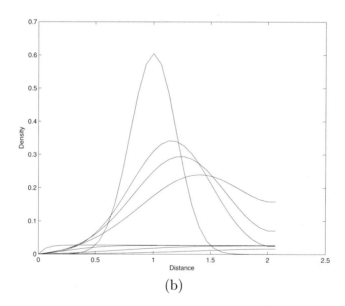

(b)

Fig. 11.3. Robust control response for one-equation uncertain-model:
(a) Contours of the response; (b) Density at different time instants

11.3 Robust Control for Two-Equation Model

In this section we formulate the control model and present robust control design for the two-equation model. This section also studies Lyapunov stability for this model and presents some simulation results. We also consider the uncertainty in the input to the system. The control design is done using a combination of both Lyapunov redesign and backstepping and the technique is referred to as robust backstepping. Let us consider the two-equation model for a 1-D corridor. The model is a higher order model or more precisely a system of two partial differential equations and consists of conservation of mass equation coupled with a second equation based on the principle of conservation of momentum. The model is given in Sect. 9.2.2 by (9.8) and (9.9) as

$$\frac{\partial \rho}{\partial t} + \frac{\partial q}{\partial x} = 0 \tag{11.25}$$

$$\frac{\partial q}{\partial t} + \frac{\partial}{\partial x}(q^2/\rho) = -\frac{\partial p}{\partial x} \tag{11.26}$$

Here ρ and q are the variables we want to control. For pedestrian evacuation the final density and flow should be equal to zero; that is $\rho(x, t_f) = 0$ and $q(x, t_f) = 0$. For this model we choose divergence of pressure $\partial p/\partial x$ as distributed control variable u thus giving us the following representation

$$\frac{\partial \rho}{\partial t} = -\frac{\partial q}{\partial x} \tag{11.27}$$

$$\frac{\partial q}{\partial t} = \bar{u} \tag{11.28}$$

where $\bar{u} = -\frac{\partial}{\partial x}(q^2/\rho) + u$ is the new control variable. Now let us consider the uncertainty in the control input for this system. We can have both matched as well as unmatched uncertainties depending upon which equation contains the uncertainty. We can have uncertainty in (11.27) where it enters the equation through the flow variable q which means that there is an error between the flow control command and the actual people flow. For this case the uncertainty is unmatched since q is the "conceptual" control input to this equation

11.3 Robust Control for Two-Equation Model

and is not the actual control input (11.28). We can also have uncertainty in the control variable $\partial p/\partial x$ in which case the uncertainty is matched and we can have both. The control design scheme in all cases is that of robust *backstepping* [27] where we do a combination of backstepping and Lyapunov redesign. Now let us discuss these cases one by one.

11.3.1 Robust Backstepping: Unmatched Uncertainty

Let us first consider the uncertainty in the input q to the nominal system (11.27). This means that there is an error in the actual flow control command q and the actual people flow of people. The control design strategy here is robust backstepping which is the combination of *backstepping* and *Lyapunov redesign*. The uncertain model is given as

$$\frac{\partial \rho}{\partial t} = -\frac{\partial}{\partial x}(q + \theta) \tag{11.29}$$

$$\frac{\partial q}{\partial t} = \bar{u} \tag{11.30}$$

where $\theta = \theta(t, u, \rho)$ is the unknown function which takes care of the disturbance in the input q to (11.27). The uncertain term for the overall system (11.30) is unmatched as it does not enter the system at the same point as the input $\bar{u}$ even though it enters equation (11.27) through the conceptual input q. As a first step we design a robust control for (11.30) using *Lyapunov redesign* technique and in the next step find the control law for the overall system (11.30) using *backstepping*.

1. *Lyapunov redesign*: The first step is to design a robust controller for (11.29) using *Lyapunov redesign* method. The goal is to design robust control law for (11.29) as

$$q = G(\rho) + v = \bar{G}(\rho) \tag{11.31}$$

such that we achieve closed loop stability and asymptotic attenuation of $\theta(t, u, \rho)$. $G(\rho)$ in (11.31) achieves closed loop stability and v asymptotically attenuates the effect of $\theta(t, u, \rho)$. The function $G(\rho)$ is designed from the feedback linearization of nominal model (11.27) for

(11.29) and v will be designed by using Lyapunov redesign method. The nominal model (11.27) is given as

$$\frac{\partial \rho}{\partial t} = -\frac{\partial q}{\partial x}$$

The stabilizing feedback controller $q = G(\rho)$ for this nominal model from (9.30) given in Chap. 9 is

$$G(\rho) = -D \int_0^x \frac{\partial^2 \rho}{\partial m^2} dm \quad (11.32)$$

which gives the conceptual nominal closed-loop system as

$$\frac{\partial \rho}{\partial t} - D\frac{\partial^2 \rho}{\partial x^2} = 0 \quad (11.33)$$

The nominal closed loop model (11.33) has an exponentially stable origin and there exists a Lyapunov function $V(t)$ given by (9.19) which satisfies $\frac{dV(t)}{dt} \leq -\beta V(t)$ with $\beta = 2DC^{-1}$. Now by using (11.32) the control law (11.31) becomes

$$q = -D \int_0^x \frac{\partial^2 \rho}{\partial m^2} dm + v \quad (11.34)$$

which gives the following closed-loop dynamics for (11.27)

$$\frac{\partial \rho}{\partial t} - D\frac{\partial^2 \rho}{\partial x^2} - \frac{\partial v}{\partial x} = 0 \quad (11.35)$$

Thus the error in the closed loop dynamics (11.33) and (11.35) due to input uncertainty is given by

$$\theta(t, \rho, u) = -\frac{\partial v}{\partial x} \quad (11.36)$$

The magnitude of this uncertain term with respect to L_2 norm is given by $\|\theta(t, \rho, u)\|_2 = -\|\partial v/\partial x\|_2$ which due to Sobolev inequality (9.21) is

$$\|\theta(t, \rho, u)\|_2 \leq \gamma \|v\|_2 \quad (11.37)$$

where $\gamma = C^{-1}$. Let us now apply control law (11.33) to the input-uncertain system (11.29) so that the closed-loop dynamics take the form

$$\frac{\partial \rho}{\partial t} - D\frac{\partial^2 \rho}{\partial x^2} - \frac{\partial v}{\partial x} - \frac{\partial \theta}{\partial x} = 0 \quad (11.38)$$

11.3 Robust Control for Two-Equation Model

Let us choose the Lyapunov function $V(t)$ for (11.29) same as before. To design v we find the rate of change of Lyapunov function as

$$\frac{\partial V(t)}{\partial t} \leq -D\|\rho\|_2^2 - \int_0^L \rho \frac{\partial v}{\partial x} dx - \int_0^L \rho \frac{\partial \theta}{\partial x} dx$$

Now let us define $\hat{\theta} = \partial \theta / \partial x$ and a new control variable $\hat{v} = \partial v / \partial x$. Therefore we have

$$\frac{\partial V(t)}{\partial t} \leq -D\|\rho\|_2^2 - \int_0^L \rho \hat{v} dx - \int_0^L \rho \hat{\theta} dx \qquad (11.39)$$

Now let us apply the Holders inequality to the second integral of (11.39). This results in the following

$$\frac{\partial V(t)}{\partial t} \leq -D\|\rho\|_2^2 - \int_0^L \rho \hat{v} dx - \|\rho\|_2 \|\hat{\theta}\|_2$$

From Sobolev inequality (9.21) we have $\|\hat{\theta}\|_2 = \|\partial \theta / \partial x\|_2 \geq C^{-1} \|\theta\|_2$, thus by using this relation in conjunction with (11.37) we get the rate of change of Lyapunov function as

$$\frac{\partial V(t)}{\partial t} \leq -D\|\rho\|_2^2 - \int_0^L \rho \hat{v} dx - C^{-1} \|\rho\|_2 \|\hat{v}\|_2 \qquad (11.40)$$

Now let us choose the control law as

$$\hat{v} = -\frac{\eta}{1-\kappa} \rho(x,t)$$

or

$$v = -\int_0^L \frac{\eta}{1-\kappa} \rho dx \qquad (11.41)$$

where $\eta(\rho, t) \geq \gamma(\rho, t)$. Using this control law in (11.41) we get

$$\frac{dV(t)}{dt} \leq -D\|\rho\|_2^2 - \|\rho\|_2^2 \left\|\frac{\eta}{1-\kappa}\right\|_2 (C^{-1} - 1)$$

which is negative semi-definite as long as $C^{-1} - 1 \geq 0$ thus giving us asymptotic stability. Therefore robust control that stabilizes the uncertain model (11.29) is given as

$$q = -D \int_0^x \frac{\partial^2 \rho}{\partial m^2} dm - \int_0^L \frac{\eta}{1-\kappa} \rho dx \qquad (11.42)$$

2. *Control by backstepping method*: The next step is to design a feedback controller for overall system (11.30) using backstepping method. The goal is to design feedback control law $\bar{u}$ to stabilize the overall system from the knowledge of Lyapunov function $V(t)$ for (11.29) and modifying it. We proceed with the control design as follows

1. First design a conceptual control law $q = \bar{G}(\rho)$ for (11.29) from the previous step to stabilize origin $\rho(x,t) = 0$. With this control law the conceptual closed loop dynamics for (11.29) are asymptotically stable and there exists a Lyapunov function to (9.19) which satisfies $\frac{dV(t)}{dt} \leq -\beta V(t)$.

2. Since q is not the actual control variable, defining the difference between q and $\bar{G}(\rho)$ by an error variable $z = q - \bar{G}(\rho)$, we get the following modified dynamics.

$$\frac{\partial \rho}{\partial t} = -\frac{\partial \bar{G}(\rho)}{\partial x} - \frac{\partial z}{\partial x} \quad (11.43)$$

$$\frac{\partial z}{\partial t} = u_n \quad (11.44)$$

where $u_n = \bar{u} - \partial \bar{G}(\rho)/\partial t$ is the new control variable.

3. Now let us modify the Lyapunov functional by adding the error term to it as in (9.34) which is given by

$$V_a(t) = V(t) + \frac{1}{2}\|z\|_2^2 = \frac{1}{2}\int_0^L |\rho|^2 \, dx + \frac{1}{2}\int_0^L |z|^2 \, dx$$

with $z = q - \bar{G}(\rho)$. The control law u_n is given by (9.35) as

$$u_n(x,t) = -Kz - z^{-1}\rho\frac{\partial z}{\partial x} \quad (11.45)$$

This control ensures that the origin of the system (11.43) and (11.44) is asymptotically stable.

Thus the final robust feedback control law for (11.29) and (11.30) is given by the following partial differential-integral equation

$$u = u_n + \frac{\partial \bar{G}(\rho)}{\partial t} + \frac{\partial}{\partial x}(q^2/\rho) \quad (11.46)$$

with $u_n(x,t)$ given by (11.45) and $\bar{G}(\rho) = -D\int_0^x \frac{\partial^2 \rho}{\partial m^2}dm - \int_0^L \frac{\eta}{1-\kappa}\rho dx$.

11.3.2 Robust Control: Matched Uncertainty

Let us consider the uncertainty in the input to the overall system (11.26) which means that there is an error between the actual value and control command for gradient of pressure $\partial p/\partial x$. By assuming the uncertainty in input u in second equation we have our input uncertain model as

$$\frac{\partial \rho}{\partial t} = -\frac{\partial q}{\partial x} \tag{11.47}$$

$$\frac{\partial q}{\partial t} = -\frac{\partial}{\partial x}(q^2/\rho) - \frac{\partial}{\partial x}(p+\theta) \tag{11.48}$$

Here the uncertainty is matched as it enters the system at the same point as actual control input u. With $|\theta(t, u, \rho)| = 0$ the nominal model for this system is given by (11.27) and (11.28). As a first step towards the control design we design feedback control for the nominal model (the system without disturbance) by applying backstepping technique. After that the robust control for the uncertain system (11.48) is designed using *Lyapunov redesign* technique. We proceed with the control design as follows

1. *Control by backstepping method*: The first step is to design a feedback controller $u = -\partial p/\partial x$ for the nominal two-equation system (11.27) and (11.28) using *backstepping* technique. The controller is given by (9.36) as

$$u = -\frac{\partial p}{\partial x} = u_n + \frac{\partial G(\rho)}{\partial t} + \frac{\partial}{\partial x}(q^2/\rho) \tag{11.49}$$

where u_n is given by (9.35) as $u_n = -Kz - z^{-1}\rho\frac{\partial z}{\partial x}$, with $z = q - G(\rho)$ and $G(\rho) = -D\int_0^x \frac{\partial^2 \rho}{\partial m^2}dm$. The nominal system (11.27) and (11.28) is asymptotically stable and there exists a Lyapunov function $V_a(t)$ given by (11.27) as

$$V_a(t) = V(t) + \tfrac{1}{2}\|z\|_2^2 = \tfrac{1}{2}\int_0^L |\rho|^2\,dx + \tfrac{1}{2}\int_0^L |z|^2\,dx \tag{11.50}$$

which satisfies $\frac{dV_a(t)}{dt} \leq -2\beta V_a(t)$ with $\beta = 2DC^{-1}$

2. *Lyapunov redesign*: The next step is to design a robust control for the uncertain model (11.47) and (11.48) using *Lyapunov redesign* method. The goal is to design feedback control law

$$\hat{u} = u + v \qquad (11.51)$$

such that we achieve closed loop stability and asymptotic attenuation of $\theta(t, u, \rho)$. u is given by (11.49) and v will be designed by using *Lyapunov redesign* method. Now with the feedback controller $\hat{u} = u + v$ or $p = \int_0^L \hat{u} dx + v$ the error in the closed-loop dynamics for the nominal and actual models due to the input uncertain term is

$$\theta(t, \rho, u) = -\frac{\partial v}{\partial x}$$

The magnitude of this uncertain term with respect to L_2 norm is given by

$$\|\theta(t, \rho, u)\|_2 = -\left\|\frac{\partial v}{\partial x}\right\|_2$$

which by using Sobolev inequality can be shown as

$$\|\theta(t, \rho, u)\|_2 \leq -\gamma \|v\|_2 \qquad (11.52)$$

where $\gamma = C^{-1}$. After applying control law 11.51 the rate of change of Lyapunov functional $V_a(t)$ is

$$\frac{dV_a(t)}{dt} \leq -D\|\rho\|_2^2 - K\|z\|_2^2 - \int_0^L \rho \frac{\partial v}{\partial x} dx - \int_0^L \rho \frac{\partial \theta}{\partial x} dx$$

or

$$\frac{dV_a(t)}{dt} = -2\beta V_a(t) - \int_0^L \rho \frac{\partial v}{\partial x} dx - \int_0^L \rho \frac{\partial \theta}{\partial x} dx \qquad (11.53)$$

Let us define $\hat{v} = \partial(v)/\partial x$ and $\hat{\theta} = \partial(\theta)/\partial x$, therefore (11.53) becomes

$$\frac{dV_a(t)}{dt} \leq -2\beta V_a(t) - \int_0^L \rho \hat{v} dx - \int_0^L \rho \hat{\theta} dx \qquad (11.54)$$

11.3 Robust Control for Two-Equation Model

By applying Holders inequality and Sobolev inequality (9.21) to the third term in the above equation and then by using (11.52) we have

$$\frac{dV_a(t)}{dt} \leq -2\beta V_a(t) - \int_0^L \rho \hat{v} dx - C^{-1} \|\rho\|_2 \|\hat{v}\|_2 \quad (11.55)$$

Let us choose the control law as before

$$\hat{v} = -\frac{\eta}{1-\kappa} \rho(x,t)$$

or

$$v = -\int_0^L \frac{\eta}{1-\kappa} \rho dx \quad (11.56)$$

where $\eta(\rho,t) \geq \gamma(\rho,t)$. Using this control law (11.54) becomes

$$\frac{dV_a(t)}{dt} \leq -2\beta V_a(t) - \|\rho\|_2^2 \left\|\frac{\eta}{1-\kappa}\right\|_2 (C^{-1} - 1)$$

Hence we have asymptotic stability as long as $(C^{-1} - 1 \geq 0)$. Thus the robust feedback control is given as $\hat{u} = u + v$ with u given by (11.49) and v is given as

$$v = -\int_0^L \frac{\eta}{1-\kappa} \rho dx \quad (11.57)$$

11.3.3 Robust Control: Both Matched and Unmatched Uncertainties

Now let us consider both matched and unmatched uncertainties for the two-equation model. This means that there is uncertainty in the input $\partial p/\partial x$ to the overall system (11.25) and (11.26) and also in the input q to (11.25) in the flow. Therefore we have our uncertain model as

$$\frac{\partial \rho}{\partial t} = -\frac{\partial}{\partial x}(q + \theta_1) \quad (11.58)$$

$$\frac{\partial q}{\partial t} = -\frac{\partial}{\partial x}\left(\frac{q^2}{\rho}\right) + \frac{\partial}{\partial x}(p + \theta_2) \quad (11.59)$$

Here the uncertainties are both matched as well as unmatched. The errors θ_1 and θ_2 are given by (11.36) and (11.52) as $\theta_1(t,\rho,u) = -\partial v_1/\partial x$ and $\theta_2(t,\rho,u) = -\partial v_2/\partial x$ respectively and are bounded as $\|\theta_1(t,\rho,u)\|_2 \leq \gamma \|v_1\|_2$ and $\|\theta_2(t,\rho,u)\|_2 \leq \gamma \|v_2\|_2$. The control design scheme will be a combination of the designs for matched and unmatched uncertainties obtained in Sects. 11.3.1 and 11.3.2. We proceed with the control design as follows

1. *Control by Lyapunov redesign*: The first step is to design a robust feedback controller $q = G(\rho) + v_1$ for (11.58) using *Lyapunov redesign* technique. The control is given by (11.43) as

$$q = -D \int_0^x \frac{\partial^2 \rho}{\partial m^2} dm - \int_0^L \frac{\eta}{1-\kappa} \rho dx \qquad (11.60)$$

with $v_1 = -\int_0^L \eta/1 - \kappa \rho dx$. The controller stabilizes (11.59) in presence of uncertainty θ_1. The Lyapunov function is given by (11.33) as $V(t) = -1/2 \|\rho\|_2^2$.

2. *Control by backstepping*: The next step is to design a controller for the nominal model (model without uncertainty θ_2) using the *backstepping* approach. The model given by (11.29) and (11.30) is the nominal model

$$\frac{\partial \rho}{\partial t} = -\frac{\partial}{\partial x}(q + \theta_1)$$

$$\frac{\partial q}{\partial t} = -\frac{\partial}{\partial x}(\frac{q^2}{\rho}) + \frac{\partial p}{\partial x} = -\frac{\partial}{\partial x}(\frac{q^2}{\rho}) + u \qquad (11.61)$$

The controller $u = -\frac{\partial p}{\partial x}$ is given by (11.46) as the following partial differential-integral equation

$$u = u_n + \frac{\partial \bar{G}(\rho)}{\partial t} + \frac{\partial}{\partial x}(q^2/\rho) \qquad (11.62)$$

The control law $u_n(x,t)$ is given by (11.45) as $u_n(x,t) = -Kz - z^{-1}\rho\frac{\partial z}{\partial x}$ with $z = q - \bar{G}(\rho)$ and $\bar{G}(\rho) = -D\int_0^x \frac{\partial^2 \rho}{\partial m^2} dm -$

$\int_0^L \frac{\eta}{1-\kappa}\rho dx$. The modified Lyapunov function for this system is

$$V_a(t) = V(t) + \tfrac{1}{2}\|z\|_2^2 = \tfrac{1}{2}\int_0^L |\rho|^2 \, dx + \tfrac{1}{2}\int_0^L |z|^2 \, dx$$

3. *Control by Lyapunov redesign*: In the final step we design the robust feedback control $\hat{u} = u+v_2$ for overall system (11.60) using Lyapunov redesign obtained in Sect. 11.3.2. The feedback control is thus given as

$$\hat{u} = u + v_2 = u - \int_0^L \frac{\eta}{1-\kappa}\rho dx \qquad (11.63)$$

The control law u is given by (11.62). The Lyapunov function for this system is same as in the previous step. Therefore we achieve asymptotic stability and disturbance attenuation with this controller.

In this chapter we designed robust feedback controllers for 1-D case. We designed controllers for uncertainty in the input. For one-equation model the uncertainty is matched and we used the method of Lyapunov redesign which achieved the attenuation of disturbance. For the two-equation model we discussed both matched as well as unmatched uncertainties and a combination of both. We used the method of *robust backstepping* which is a combination of *Lyapunov redesign* and *backstepping*. In all the controllers we achieved both asymptotic stability and disturbance attenuation.

11.4 Computer Code

In this section the code used for the above simulations is included. The code is in matlab and consists of m-file named robust_1d

11.4.1 robust_1d

```
L=2; %%%length
T=1;  %%%%total time
n=30; %%% space samples
d_x=L/n;  %%%%% spacing in x-direction
```

```
d_t=0.002;  %%%%%  spacing in time
N=T/d_t;  %%%%% time samples
V=4;
D=0.9;
rm=10;
h=d_x;
g=d_t;
s=(g/h);
f=(g/(h^2));
K=g/(h^2);
x=[0:d_x:L];
t1=[0:d_t:T];

std=L/10;meanx=1; Row_0 = ((10/(std*sqrt(2*pi)))/33)
*exp(-(x-meanx).^2./(2*std^2));
r(1,1:n+1)=Row_0(1:n+1);
V2=20;      %%%%%% Initial Condition

for t=1:N; % Total time = t* d_t
for j=2:n+1;
if j==2;r(t,j-1)=0*r(t,j);end
if j==n+1 ; r(t,j+1)=1*r(t,j);end
r(t+1,j)=0.5*(r(t,j+1)+r(t,j-1))-(s/2)*V....
*(r(t,j+1)-r(t,j-1));
if t > 49 && t < 81
 r(t+1,j)=0.5*(r(t,j+1)
         +r(t,j-1))-(s/2)*(V-V2)....%%%step noise
*(r(t,j+1)-r(t,j-1));
w(t+1,j)=r(t+1,j)*(1-r(t+1,j)./rm);
d_w(t+1,j)=(1/h)*(w(t+1,j)-w(t+1,j-1));
v(t+1,j)=-(w(t+1,j).*w(t+1,j));%./d_w(t+1,j);
 r(t+1,j)=0.5*(r(t,j+1)
         +r(t,j-1))-(s/2)*(V-V2)....%%%step noise
*(r(t,j+1)-r(t,j-1))+v(t+1,j);

end
end
end
```

11.4 Computer Code

```
x2=[0:d_x:L+d_x];
figure(1)
plot(x2,r(1,:),'g');
hold on
plot(x2,r(20,:),'b');
plot(x2,r(30,:),'c');
plot(x2,r(50,:),'r');
plot(x2,r(80,:),'k');
plot(x2,r(100,:),'m');
plot(x2,r(150,:),'b');
plot(x2,r(250,:),'y');
hold off
xlabel('Distance')
ylabel('Density')
figure(2)
[c,h1] = contour(x2,t1,r); clabel(c,h1,'manual');
xlabel('Distance')
ylabel('Time')
```

Bibliography

[1] S. Al-nasur and P. Kachroo. A microscopic-to-macroscopic crowd dynamic model. In *9th International IEEE Conference on ITSC*, pp. 606–611, 2006.

[2] S. A. H AlGadhi and H. Mahmassani. Simulation of crowd behavior and movement: Fundamental relations and applications. *Transportation Research Record*, 1320:260–268, 1991.

[3] J. D. Anderson. *Fundamentals of Aerodynamics*. McGraw-Hill Science, 2001.

[4] A. Aw and M. Rascle. Reconstruction of 'second order' models of traffic flow. *SIMA Journal of Applied Mathematics*, 60: 916–938, 2000.

[5] D. L. Bakuli and J. M. Smith. Resource allocation in state-dependent emergency evacuation networks. *European Journal of Operations Research*, 89:543–555, 1996.

[6] M. Bando. Dynamical model of traffic congestion and numerical simulation. *Physical Review E*, 51:1035–1042, 1995.

[7] B. Barrett, B. Ran, and R. Pillai. Developing a dynamic traffic management modeling framework for hurricane evacuation. In *Transportation Research Record*, 1733:115–121, Washington, DC, 2003.

[8] A. Belleni-Morante. *Applied Semigroups and Evolution Equations*. Clarendon Press, Oxford, 1979.

[9] A. Belleni-Morante. *A Concise Guide to Semigroups and Evolution Equations*. World Scientific, 1994.

[10] V. J. Blue and J. L. Adler. Emergent fundamental pedestrian flows from cellular automata microsimulation. *Trasportation Research Record*, 1644:29–36, 1998.

[11] Transportation Research Bord. Highway capacity manual. Special Report 204 TRB, 1985.

[12] P. H. L. Bovy and E. Stern. *Route Choice: Wayfinding in Transport Networks*. Kulwer Academic Publishers, Dordrecht, 1990.

[13] A. Bressan. An ill posed Cauchy problem for a hyperbolic system in two space dimensions. Rend. Sem. Mat. Univ. Pasova, 2003.

[14] J. A. Burns and S. Kang. A control problem for Burgers equation with bounded input/output. *Nonlinear Dynamics*, 2:235–262, 1991.

[15] C. Burstedde, A. Klauck, and J. Zittartz. Simulation of pedesrtian dynamics using a two-dimensional cellular automaton. *Physica A: Statistical Mechanics and its Applications*, 295(3–4):507–525, 2001.

[16] L. Chalmet, R. L. Fransis, and P. B. Saunders. Network models for building evacuation. *Management Science*, 28(1):86–105, 1982.

[17] R. E. Chandler, R. Herman, and E. W. Montroll. Traffic dynamics; studies in car following. *Operations Research*, 6:165–184, 1958.

[18] P. D. Christofides. *Nonlinear and Robust Control of PDE Systems*. Birkhauser, 2000.

[19] M. J. Corless and G. Leitmann. Continuous state feedback guaranteeing uniform ultimate boundedness for uncertain dynamic systems. *IEEE Transactions on Automatic Control*, 26:1139–1144, 1981.

[20] R. Cote and G. E. Harrington. Life safety code handbook. Technical report, NFPA, Avon, MA, 1987.

Bibliography

[21] T. J. Cova and J. P. Johnson. A network flow model for lane based evacuation routing. *Transportation Research Part A*, 37:579–604, 2003.

[22] M. N. Tali-Manaarn, D. Dochain, and J. P. Babary. Modeling and adaptive control of nonlinear distributed parameter bioreactors via orthogonal collocation. *Automatica*, 68:873–883, 1992.

[23] C. Daganzo. Requiem for second-order fluid approximation to traffic flow. *Transportation Research Part B*, 29B(4):277–286, 1995.

[24] D. Emmanuele. *Partial Differential Equation*. Birkhauser, Boston, 1995.

[25] L. C. Evans. *Partial Differential Equations*. American Mathematical Society, 1998.

[26] S. J. Farlow. *Partial Differential Equations for Scientists and Engineers*. John Wiley, Canada, 1982.

[27] R. A. Freeman and P. V. Kokotovic. *Robust Nonlinear Control Design: State-Space and Lyapunov Techniques*. Birkhauser, 1996.

[28] A. Friedman. *Partial Differential Equations*. Holt, Rinehart and Winston, New York, 1976.

[29] J. J. Fruin. Designing for pedestrians: A level of service concept. *Highway Research Record*, 355:1–15, 1971.

[30] J. J. Fruin. *Pedestrian Planning and Design*. Metropolitan Association of Urban Designers and Enviromental Planners, Inc, New York, 1971.

[31] P. G. Gipps. Simulation of pedestrian traffic in buildings. Technical Report, University of Karlsruhe, 1986.

[32] P. G. Gips and B. Marksjo. A micro-simulation model for pedestrian flows. *Mathmatics and Computer in Simulation*, 27:95–105, 1985.

[33] H. Greenberg. An analysis of traffic flow. *Operationa Research*, 7:78–85, 1959.

[34] B. D. Greenshields. A study in highway capacity. *Highway Research Board*, 14:458, 1935.

[35] D. A. Griffith. The critical problem of hurricane evacuation and alternative solutions. In *CAITR Paper 15th Conference of Australian Institutes of Trasport Research*, 12:990–994, 1982.

[36] P. K. Gundepudi and J .C. Friedly. Velocity control of hyperbolic partial differential equation systems with single characteristic variable. *Chemical Engineering Science*, 53:4055–4072, 1998.

[37] R. Haberman. *Mathematical Models*. Prentice-Hall, New Jersey, 1977.

[38] D. Helbing. A mathmatical model for the behavior of pedestrians. *Behavioral Science*, 36:298–310, 1991.

[39] D. Helbing. A fluid-dynamic model for the movement of pedestrians. *Complex Systems*, 6:391–415, 1992.

[40] D. Helbing. Computer simulation of pedestrian dynamics and trail formation. *Evolution of Natural Science*, 230:229–234, 1994.

[41] D. Helbing, I. Farkas, and T. Viscek. Simulating dynamic features of escape panic. *Nature*, 407:487–490, 2000.

[42] D. Helbing, I. J. Farkas, P. Molnor, and T. Vicsek. Simulation of pedestrian crowd in normal and evacuation situations. *Pedestrian and Evacuation Dynamics Journal*, 21–58, 2002.

[43] D. Helbing and P. Molnar. Social force model forpedestrian dynamics. *Physical Review E*, 51:4282–4286, 1995.

[44] D. Helbing and P. Molnar. Self-Organization Phenomena in Pedestrian Crowds. *ArXiv Condensed Matter e-prints*, 569–577, June 1997.

[45] D. Helbing and T. Vicsek. Optimal self-organization. *New Journal of Physics*, 1:13.1–13.17, 1999.

[46] L. F. Henderson. On the fluid mechanic of human crowd motion. *Transportation Research*, 8:509–515, 1974.

[47] M. R. Hill. *Patial Structure and Decision-Making of Pedestrian Route Selection through an Urban Environment*. PhD thesis, University Microfilms International, 1982.

[48] E. N. Holland. A generalized stability criterion for motorway traffic. *Transportation Research Part B*, 32:141–154(14), 1998.

[49] S. P. Hoogendoorn. Pedestrian flow behavior: Theory and applications. Technical Report, Transportation and Traffic Engineering Section, Delft University of Technology, 2001.

[50] S. P. Hoogendoorn and P. H. L. Bovy. Pedestrian route-choice and activity scheduling theory and models. *Transportation Research Part B*, 38:169–190, 2004.

[51] R. L. Hughes. A continuum theory for the flow of pedestrians. *Tranportation Research Part B*, 36:507–535, 2002.

[52] R. L. Hughes. The flow of human crowds. *Annual Review of Fluid Machanics*, 35:169–182, 2003.

[53] P. Kachroo and K. Kumar. System dynamics and feedback control design formulations for real time ramp metering. *Transactions of the SDPS, Journal of Integrated Design and Processes Science*, 4(1):37–54, 2000.

[54] P. Kachroo and K. Ozaby. Solution to the user equilibrium dynamic traffic routing problem using feedback linearization. *Transportation Research Part B*, 32(5):343–366, 1998.

[55] P. Kachroo and K. Ozaby. *Feedback Control Theory for Ramp Metering in Intelligent Transportation Systems*. Kulwer, 2004.

[56] P. Kachroo. Microprocessor controlled small scale vehicles for experiments in automated highway systems. *The Korean Transport Policy Review*, 4(3):145–178, 1997.

[57] P. Kachroo and K. Ozbay. *Feedback Control Theory For Dynamic Traffic Assignment.* Springer-Verlag, 1999.

[58] J. Kevorkian. *Partial Differential Equation.* Springer, New York, 2nd edition, 2000.

[59] H. K. Khalil. *Nonlinear Systems.* Prentice Hall, 1996.

[60] A. Kirchner, H. Kulpfel, K. Nishinari, A. Schadschneider, and M. Schreckenberg. Discretization effects and the influence of walking speed in cellular automata models for pedestrian dynamics. *Journal of Statistical Mechanics-Theory and Experiment: Art. No. P10011*, October 2004.

[61] A. Kirchner, K. Nishinari, and A. Schadschneider. Friction effects and clogging in a cellular automaton model for pedestrian dynamics. *Physical Review E*, 67:056122, 2003.

[62] A. Kirchner and A. Schadschneider. Simulation of evacuation process using a bionics inspired cellular automaton model for pedestrian dynamics. *Physica A: Statistical Mechanics and its Applications*, 312(1–2):260–276, 2002.

[63] D. E. Kirk. *Optimal Control Theory: An Introduction.* Prentice-Hall, Inc., Englewood Cliffs, 1970.

[64] D. Kröner. *Numerical Schemes for Conservation Laws.* Wiley, Teubner, 1997.

[65] S. N. Kruzkov. First order quasi-linear equations in several independent variables. *Mathematics of USSR Sbornik*, 10: 217–243, 1970.

[66] P. D Lax. Hyperbolic systems of conservation laws and mathmatical theory of shock waves. In *SIMA Regional Conference Series in Applied Mathematics*, 11, 1972.

[67] R. J. Leveque. *Finite Volume Methods for Hyperbolic Problems.* Cambridge University Press, UK, 2002.

[68] M. J. Lighthill and G. B. Whitham. On kinematic waves. i:flow movement in long rivers. ii:a theory of traffic on long crowded roods. In *Proc. Royal Soc.*, A229:281–345, 1955.

[69] Richard Liska and Burton Wendroff. 2nd shallow water equations by composite schemes. *International Journal for Numerical Methods in Fluids*, 30(4):461–479, 1997.

[70] G. G. Lovas. Modeling and simulation of pedestrian traffic flow. *Tranportation Research*, 28B:429–443, 1994.

[71] G. G. Lovas. On the importance of building evacuation system components. *IEEE Transactions On Engineering Management*, 45(2):181–191, 1998.

[72] S. Mamada, K. Makino, and S. Fujishige. Evacuation probllems and dynamic network flows. In *SICE Annual Conference*, Sapporo, Japan, August 2004.

[73] A. C. May. *Traffic Flow Fundemental*. Prentice Hall, New Jersey, 1990.

[74] T. Musha and H. Higuchi. Traffic current fluction and the burgers. *Japanese Jounal of Applied Physics*, 17:811–816, 1978.

[75] K. Nagel. Partical hopping models and traffic flow theory. *Physical Review E*, 53:4655–4672, 1996.

[76] H. Nijmeijer and A. J. van der Schaft. *Nonlinear Dynamical Control Systems*. Springer-Verlag, New York, 1990.

[77] J. Ockendon, S. Howison, A. Lacey, and A. Movchan. *Applied Partial Differential Equations*. Oxford, New York, revised edition, 2003.

[78] S. Okazaki. A study of pedestrian movement in architectural space, part 1: Pedestrian movement by the application on magnetic models. *Trans. of A.I.J.*, 283:111–119, 1979.

[79] S. Okazaki and S. Matsushita. A study of simulation model for pedestrian movement with evacuation and queuing. In *Proceeding of the International Confference on Engineering for Crowd Safety*, 12:271–280, 1993.

[80] T. Okumura, N. Takasu, S. Ishimatsu, S. Miyanoki, A. Mitsuhashi, K. Kumada, K. Tanaka, and S. Hinohara. Report

on 640 Victims of the Tokyo Subway Sarin Attack. *Annals of Emergency Medicine*, 28(2):129–135, 1996.

[81] M. Papageorgiou. *Applications of Automatic Control Concepts to Traffic Flow Modelling and Control.* Springer-Verlag, 1983.

[82] H. J. Payne. Models of freeway traffic and control. In *Math. Models Publ. Sys. Simul. Council Proc.*, 28:51–61, 1971.

[83] A. Pazy. *Semigrous of Linear Operators and Applications to Partial Differential Equations.* Springer-Verlag, New York, 1983.

[84] M. Rascle. An improved macroscopic model of traffic flow: Derivation and lonks with the lighthill-whitham model. *Mathematical and Computer Modelling*, 35:581–590, 2002.

[85] W. H. Ray. *Advanced Process Control.* McGraw-Hill, New York, 1981.

[86] M. Renardy and R. Rogers. *Introduction to Partial Differential Equation.* Springer, New York, 2nd edition, 2004.

[87] P. I. Richards. Shockwaves on the highway. *Operationa Research*, 4:42–51, 1956.

[88] W. J. Rugh. *Linear Systems Theory.* Prentice Hall, 1996.

[89] A. Schadschneider. Cellular automata approach to pedestrian dynamics-theory. In M. Schreckenberg and S. Sharma, editors, *Pedestrian and Evacuation Dynamics*, pp. 75–86. Springer, 2002.

[90] J.-J. Slotine and W. Li. *Applied Nonlinear Control.* Prentice Hall, New Jersey, 1991.

[91] R. A. Smith and J. F. Dickie (Eds.). Engineering for crowd safety. Elsevier, Amsterdam, p. 442, 1993.

[92] F. Southworth, S.-M. Chin, and P. D. Cheng. A telemetric monitoring and analysis system for use during large scale population evacuations. In *Road Traffic Monitoring, Second International Conference*, pp. 99–103, 1989.

[93] Y. Sugiyama, A. Nakayama, and K. Hasebe. 2-dimensional optimal velocity methods for granular flow and pedestrian dynamics. *Pedestrian and Evacuation Dynamics*, pp. 155–160, 2002.

[94] P. A. Thompson and E. W. Marchant. A computer model the evacuation of large building populations. *Fire and Safety Journal*, 24:131–148, 1995.

[95] P. A. Thompson and E. W. Marchant. Testing and application of the computer model 'simulex'. *Fire and Safty Journal*, 24:149–166, 1995.

[96] R. L. Tobin, T. L. Friesz, and B. W. Wie. Dynamic network traffic assignment considered as a continuous time optimal control problem. *Operations Research*, 37:893–901, 1989.

[97] E. F. Toro. *Riemann Solvers and Numerical Methods for Fluid Dynamics*. Springer, Germany, 2nd edition, 1999.

[98] P. Varaiya. Smart cars on smart roads: Problems of control. *IEEE Transactions On Automatic Control*, 38(2), 1993.

[99] S. Wadoo and P. Kachroo. Feedback control design and stability analysis of one dimensional evacuation system. In *Proceedings of the IEEE Intelligent Transportation Systems Conference*, Canada, pp. 618–623, 2006.

[100] S. Wadoo and P. Kachroo. Feedback control design and stability analysis of two dimensional evacuation system. In *Proceedings of the IEEE Intelligent Transportation Systems Conference*, Canada, pp. 1108–1113, 2006.

[101] J. M. Watts. Computer models for evacuation analysis. *Fire and Safety Journal*, 12:1237–245, 1987.

[102] G. B. Whitham. *Linear and Nonlinear Waves*. John Wiley, New York, 1974.

[103] M. R. Wigan. Why should we worry about pedestrians? In *CAITR Paper 15th Conference of Australian Institutes of Trasport Research*, p. 12, December 1993.

[104] H. M. Zhang. A theory of nonequilibrium traffic flow. *Transportation Research Part B*, 32:485–498, 1998.

[105] H. M. Zhang. A non-equilibrium traffic model deviod of gas-like behaviour. *Transportation Research Part B*, 36B:275–290, 2002.

Index

Anisotropic, 7, 33, 40

Backstepping
    distributed, 175
    one dimensional, 175
    two dimensional, 199

Calculus of Variation, 142, 144
Car-following
    2-D, 45
Cellular Automata, 122, 141
Co-state, 144–146, 149, 151, 152
Conservation
    1-D differential form, 11
    1-D integral form, 10, 11
    law, 9
    mass, 164
    momentum, 165
Conservation of momentum
    2-D, 197
Conservation of momentum
    one dimensional, 166
Constraints
    State, 125, 129
Continuity equation
    one dimensional, 164
    two dimensional, 191
Continuum, 122, 141
Control
    feedback, 166
    saturation, 172

Control Gain, 126
Control Parameter, 123, 128, 130
Convective derivative, 18

Density, 8
    microscopic, 7
Determinant, 127
Discharge, 123, 124
Discretization
    of Control, 129
    of pde, 122, 124
    of Control, 131, 133
Distributed feedback
    one dimensional, 167
    two dimensional, 192
Dynamic Traffic Assignment, 122, 141

Entropy
    2-D scalar, 34
    lax, 31

Feedback
    Control, 141
    Linearization, 125, 126, 131
    Control, 121, 126, 129, 131–133
Feedback Control
    one dimensional, 166
    two dimensional, 193

Flow
    1-D, 8
Free Flow Velocity, 124, 128, 131–133, 143

Greenshield's Model, 124, 125, 152
Greenshields
    one dimensional, 165
    two dimensional, 192

Hamiltonian, 144–146, 151
Hilbert space
    infinite dimensional, 164
Hyperbolic
    strictly, 23
    system, 35
        strictly, 35

Isotropic, 17, 33, 36, 40

Jam Density, 124, 131
Jam density, 8, 10, 45
Jamming, 122, 131

Lagrangian
    AR model, 18
Leibnz rule, 173
Linear System, 125, 126, 128
Linearize, 53
    first system mode, 55
    scalar crowd model, 53
    second system model, 57
    third system model, 58
Link Level, 122
Lyapunov
    functionl, 169
    stability, 168
Lyapunov Stability, 168, 193

Macroscopic Traffic Modeling, 122, 141
Microscopic Traffic Modeling, 141

Negative Definate, 126
Network Level, 122
Norm, 169

One Equation Model, 164
    Control, 166
    Control Model
        one dimensional, 167
    one dimensional, 165
One equation model
    Control Model
        two dimensional, 192
Operator
    bouded linear, 168
    nonlinear, 167, 168
Optimal Control, 141, 159

Positive Definate, 127

Saturation of Control, 128
Semigroup
    continuous, 168
Sobolev Inequality, 170
Sobolev spaces, 165
Solution
    admissible, 31
    discontinuous, 27
    fan, 27
    rarefaction, 27, 30
    shock, 26
    weak in 1-D, 27
    weak in 2-D scalar, 34
Space
    normed, 168
Stability, 126, 128, 168

Index

    Asymptotic, 128
    asymptotic, 169
    exponential, 169
Stabilization
    one dimensional, 167
State Equations, 125, 128, 142–144, 152
    Linear, 125
    Non-Linear, 125
State Variable, 122, 125, 126

State-Space, 125

Taylor series, 53
Two Equation Model, 165
    Control
        one dimensional, 175
        two dimensional, 197
    one dimensional, 166

Youngs Inequality, 175

Printing: Krips bv, Meppel, The Netherlands
Binding: Stürtz, Würzburg, Germany